CURRENT GOOD MANUFACTURING PRACTICES FOOD PLANT SANITATION

SECOND EDITION

by

Wilbur A. Gould, Ph.D.
Food Industries Consultant,
Executive Director
Mid-America Food Processors
Association,
and,
Emeritus Professor
Food Processing & Technology
The Ohio State University

CGMP'S/Food Plant Sanitation

CTI PUBLICATIONS, INC.
Baltimore, Maryland USA

CTI Publications, Inc.
2 Oakway Road, Timonium, Maryland 21093-4247 USA
Printed in the United States of America

ISBN Numbers are as follows: 0-930027-21-3

Library of Congress Cataloging - in - Publication Data

Gould, Wilbur A., 1920-
Current Good Manufacturing Practices/Food Plant Sanitation / by Wilbur A. Gould - 2nd Edition.
p. cm.
Title on verso of t.p.: CGMP's/Food plant sanitation.
Rev. Ed. of CGMP's/Food Plant Sanitation. C1900.
Includes bibliographical references and index.
ISBN 0-930027-21-3
1. Food processing plants--Sanitation. 2.Food industry and trade--Sanitation. 3. Food Handling. I. Gould, Wilbur A., 1920- CGMP's/Food Plant Sanitation. II. Title. III. Title: CGMP's Food Plant Sanitation.
TP373.6.G68 1994
664'.02--dc20 93-41592 CIP

CTI Publications, Inc.

2 Oakway Road, Timonium, Maryland 21093-4247 USA

Voice 410-308-2080   FAX 410-308-2079

Other Titles From CTI Publications

PREFACE
First Edition

The food industry has one common goal, that is, to prepare, process, package, and preserve high quality foods that make for repeat sales. Consumers make the decision to single out products at the market place and purchase that product if it satisfies their needs. Once consumers find a product that satisfies their need, they will tend to continue to purchase that product until the product does not meet that need. Generally the first prerequisite to change a product in the eye of the consumers comes about because the product does not satisfy their standards of quality. Consumers standards of quality are based first on things they can visually see in that product. If a product has defects present or has signs of use of defective materials detectable by smell or flavor, consumers will change products and the ultimate loser is the original processor. Therefore, it behooves the food industry to maintain and constantly improve the products they manufacture for continued repeat business.

This book was developed to help the personnel responsible for food plant sanitation understand the areas of major concern in the manufacturing of high quality foods. The book, also, presents the pertinent parts of the Food Laws and Regulations to guide the industry in the understanding of the Current Good Manufacturing Practices and how they apply to food plant sanitation. Practical control and cleaning procedures are discussed as well as the problem areas to be concerned with keeping a food plant clean. Emphasis is given to the development of the training of employees and their role in the processing of clean, defect free, and safe foods.

Every effort has been made to present the material in this text in practical terms for the help of all personnel from the worker to the sanitation crew through to management to better understand "the need for", "the why of", and "the how to" aspects of food plant sanitation.

Wilbur A. Gould

ACKNOWLEDGEMENTS
First Edition

I am deeply indebted to my former students, both undergraduate and graduate students, and my former colleagues at The Ohio State University for their interest in the development of my Sanitation Course and their participation therein. Also, I am deeply indebted to the many industry personnel who have taken my Good Manufacturing Practices/Food Plant Sanitation Seminars and Workshops over the past 25 years. Their comments and suggestions have been most helpful and an inspiration to put my handouts and materials together in book form. Further, I sincerely thank the many supply firms for their inputs with photos, visuals, suggestions, and text quotes.

The selected use of material from the Food and Drug Administration, the National Food Processors Association, and the Food Processors Institute as well as many publications in the field of food plant sanitation has been most helpful and my sincere thanks to all concerned.

I am most appreciative of the assistance of Ronald W. Gould in his critique of this book and his valuable suggestions throughout the writing of same and to Jessie Gould for her constant help and valuable suggestions. Finally, my sincere thanks to Art Judge, II and Randy Gerstmyer for their valuable help and aid in completing this book and presenting it to the industry.

Wilbur A. Gould

PREFACE
Second Edition

The first edition of this book has found wide use as a textbook, training manual, and reference book for many in the food industries. This second edition has been completely up-dated, revised where necessary, and added to expand on the whole subject of food plant sanitation and Current Good Manufacturing Practices.

However, if one takes the time to read the WARNING LETTERS sent to segments of the food industries, one must conclude that we have some serious problems within this great industry that some food processors need to address promptly. As an example, in a recent letter one food firm was warned of employees wearing tank top shirts; ineffective hair restraints; drinking beverage not confined to areas where food, equipment, and utensils may be exposed to contamination; safety type light fixtures not provided, ventilation fan not covered with screen; two-inch gap along bottom of door; unclosed doors; rusty and dirty ladders holding utensils; food contact surfaces not clean or sanitized; hose from floor placed in food, etc. There is no excuse for these practices, nor can the food industry afford to let them exist. Good Manufacturing Practices and food plant sanitation are not tough subjects to comprehend. They can be accomplished with ease, but first they must come from within, that is management to every employee. We must develop pride in what we do, when we do it, where we do it, how we do it, why we do it, and who does it. It is everybody's job and we all must arise above any potential suspicion. The CGMP's are our guide, but we all know we can do much better.

Two new chapters have been added to this edition. The first on Metal Detection and the second on Sanitation Training Programs for Food Plant Personnel. Both of these additions are important as they cover two aspects of food plant sanitation not covered in the first edition. In the chapter on Metal Detection, information on the use of metallic band aid is included. In my

CGMP training classes several "students" have asked for this information and where the metal band aides may be obtained. I have listed a source, but I am sure there are many other sources. One aspect of metal detection of interest is that a straight metal detection may not be the best piece of equipment. Rather, the metal detection equipment should not be only magnetic, but it should be an electronic metal detection device as this device picks up more than just ferrous substances.

In the chapter on Training Programs, I have used some examples of teaching methods used in my 40 years of teaching at The Ohio State University. Effective teaching is not just reciting or reading, it is using practical examples and illustrations. My working with industry has given me an opportunity to gain much knowledge that I have found most useful in my teaching program.

I have, also, up-dated the Regulations. However, no segment of the food industry should wait for changes in the regulations to comply. Industry must stay out in front of any regulation and they must lead rather than follow. Leaders in this great food industry should set the example and really lead any government regulator. Industry has the knowledge and those who strive for excellence should share this information with their competitors. It is the safety of our food supply and the satisfaction of our enthusiastic customers that we must satisfy. If we in this great food industry give the customer what they expect all the time, there will be no need to worry about laws and regulations and what to do when the inspector comes. We can and we will as an industry lead the way with good sanitary practices that lead to improved good manufacturing practices.

Wilbur A. Gould

ACKNOWLEDGEMENTS
Second Edition

My thanks to the many supply firms to the food industries for their complete cooperation in providing me with information, pictures, technical literature, and help in my revision of this book. Special thanks to John Swartz of Conney Safety Products, Madison, Wisconsin, for their help with the detectable adhesive strips; to Meritech, Inc., Englewood, Colorado, for their technical info and all on hand washing systems; to CSI Supply Company, Inc., Columbus, Ohio, for their info on Kemlite, Glasbord, and Sanigrid fiberglass ceiling grids; and Donald R. Hannes of Cintex of America, Kenosha, Wisconsin and Peter R. Ledger, Yamato Lock Inspections Systems, Fitchburg, Massachusetts, for their information, photos and all on Electronic Metal Detection.

I wish to acknowledge with a big THANK YOU to CTI Publications, particularly Art Judge II, Randy Gerstmyer, Nancy Gerstmyer and others for their encouragement, interest, and full cooperation with this second edition. I also thank Ronald Gould, Jessie Gould and others for their help and assistance with this edition. Finally, I wish to say thanks to those supply firms for their contributions, interest, and help and to the many "students" in the food industry for their active interest in this subject. I trust my efforts will be helpful and with this second edition will find even much wider use and help to all concerned.

Wilbur A. Gould

This Copy Of

CGMP'S/Food Plant Sanitation

2nd Edition

Belongs to:

____

ABOUT THE AUTHOR

Wilbur A. Gould was reared on a farm in Northern New Hampshire. He received his Bachelor of Science degree from the University of New Hampshire in Horticulture-Plant Breeding. He started his graduate work at Michigan State University prior to service in the U.S. Navy during World War II. After military service, he completed his Master of Science and Ph.D degrees at The Ohio State University.

Dr. Gould retired from The Ohio State University after 39 years on the faculty as Professor of Food Processing and Technology. He taught 9 courses during his tenure and advised over 900 undergraduate students, 131 Master of Science Students and 76 Doctoral students. His major research interests were in Vegetable Processing and Technology and Snack Food Manufacture and Quality Assurance. He has authored some 83 referred journal research publications, over 200 Food Trade articles, and 10 books.

Dr. Gould is a Member of Phi Kappa Phi, Phi Sigma, Phi Tau Sigma, Sigma Xi, Gamma Sigma Delta (Award of Merit in 1984), Alpha Gamma Rho, Institute of Food Technologists (Fellow in 1982), and American Society of Horticultural Science (Distinguished Graduate Teaching Award in 1985).

The following are some of the recognitions that Dr. Gould has received: The Ohio State University Distinguished Leadership to Students Award in 1963 and a Certificate of Recognition Award in 1986; Ohio Food Processors H.D. Brown Person of Year Award in 1971; Ohio Food Processors Association Tomato Achievement Award in 1985; Ozark Food Processors Association Outstanding Professional Leadership Award in 1978; 49er's

Service Award in 1979; Food Processing Machinery and Supplies Association Leadership and Service Award in 1988; Ohio Agricultural Hall of Fame in 1989, an Honorary Life Membership in Potato Association of America in 1990, and was the Institute of Food Technologist's 1993 Nicholas Appert Award Medalist.

Dr. Gould presently serves as Executive Director of Mid-America Food Processors Association, Food Technology Consultant to the Snack Food Association, Secretary-Treasurer of The Guard Society, and Consultant to the Food Industries.

Dr. Gould's philosophy is to tell it as he sees it, be short and get right to the point.

Table of Contents

CGMP'S/FOOD PLANT SANITATION

by Wilbur A. Gould

CHAPTER 1

THE IMPLICATIONS OF CURRENT GOOD MANUFACTURING PRACTICES (CGMP'S) AND FOOD PLANT SANITATION

In the production of foods for human consumption, the art and practice of modern food plant sanitation principles and the adoption of current good manufacturing practices are mandatory for public acceptance of your products. These principles and practices are constantly being up-graded as man's knowledge increases with the ultimate production of higher quality food for all concerned.

During the past 100 years, great changes have taken place in the concepts and practices of sanitation. In the early days, gross sources of contamination included insects, rodents, sticks, stones, straw, wood, sand and dirt. Man has learned that most of these sources of contamination can be eliminated by maintaining physical cleanliness in and around the food plant. Looking beyond these gross sources of contamination, today one is concerned with microbiological sources of contamination. Microorganisms may come from people working in the plant or handling the food, from raw products, *from the food materials or ingredients* or from lack of proper cleaning of the food plant equipment or the food plant. A third problem area is potential chemical contamination. This source of contamination can be controlled by proper usage of pesticides. That is, using the right pesticides on the right crop, at the right time, and in the right amount. This same principle applies to chemicals used in the manufacturing of the food, that is, the proper use and application of food additives and/or chemicals used in the cleaning and sanitizing of the food plant.

After the passage of the Food Drug and Cosmetic Act of 1938 many changes in food plant sanitation have been adopted. Criteria for contamination and adulteration have brought to the forefront the need for better plant sanitation. Much more has happened during the past 15 years since the enactment of the GMP's and the Current GMP's. These GMP's have elucidated refinements in equipment, plant food handling, personnel, etc. that have aided food plant management in better understanding the public health concerns of producing clean, safe, and wholesome foods. Today we know how to do it, but we don't always practice the knowledge that we have.

Nearly every person working in the food industries is there to produce, process, pack, and distribute foods for human consumption. Each person has a legal and a moral obligation to perform all unit operations in clean surroundings and with due regards to the basic principles of sanitation. Everyone should understand and practice the motto of the National Sanitation Foundation.

FIGURE 1.1 – The manager must make the ultimate decision for building the sanitary program

This motto is as follows:

> *Sanitation is a way of life. It is the quality of living that is expressed in the clean home , the clean farm, the clean business and industry, the clean neighborhood, the clean community. Being a way of life it must come from within the people; it is nourished by knowledge and grows as an obligation and an ideal in human relations. (Taken from National Sanitation Institute)*

Thus, sanitation can be simply defined as the production, manufacture and distribution of clean and wholesome foods by people. People are the one most important single variable in food plant sanitation. They need training and an understanding of "the why", "the how", "the when", the what", the who", and "the where" of food plant sanitation. Once they understand the principles of food plant sanitation and the specific applications to their area of responsibilities, they need to be empowered to act, that is, make the appropriate decisions to maintain a clean factory, clean equipment, clean procedures, etc. to manufacture pure and wholesome foods. All employees must be held accountable to do what must be done all times to assure safe high quality products. People are the key to food plant sanitation.

From a practical standpoint, the ultimate in food plant sanitation is to attain the highest goal in quality of the products being produced and doing this at proper and reasonable total costs. Foods for human consumption must be safe, wholesome and appealing to the consumer. What was good enough yesterday, may not be good enough today.

Every food firm must have a goal for every manufactured product. This goal may be defined in terms of specifications understood by purchasing, processing and distribution. Specifications are a means of communication between management, plant employees, vendors, and the ultimate consumers. Without specifications, sanitation standards are vague and only concepts in the eyes of the consumers. Specifications help everyone better understand what is wanted. If specifications are properly written and understood, they provide all with facts that can be followed to produce acceptable products in a clean plant from clean raw materials and ingredients, on clean equipment, by knowledgeable people in clean environments.

Sanitation is every person's job in a food plant. Sanitation should be a part of the everyday policy of the food firm. Only through the individual efforts of each person working in a food plant can a firm expect to maintain the respect that your products demand by the consumer. Sanitation is a responsibility that every person handling or working with food must consistently fulfill.

If properly practiced, sanitation should remove the worry about the spreading of communicable diseases or potential food poisoning. Further, if sanitation is properly maintained, a product free of contamination will be produced and waste and spoilage will be eliminated. All of these are moral obligations.

From a legal obligation standpoint, the FD&C Act states in Section 402 (a) 3 that a food shall be deemed to be adulterated "if it consists in whole or in part of any filthy, putrid, or decomposed substance, or if it is otherwise unfit for food" and in Section 402 (a) 4 it states: "if it has been prepared, packed, or held under insanitary conditions whereby it may become contaminated with filth, or whereby it may have been rendered injurious to health". These definitions of adulteration and contamination are, in great part, the essence of any good manufacturing practice and/or food plant sanitation program.

The value of a planned sanitation program utilizing good manufacturing practices includes the following:

1. A better product to meet the competition's demands and consumer's expectations.
2. A more efficient food plant operation.
3. Greater employee productivity.
4. Fewer food plant employee accidents.
5. Fewer consumer complaints.

Thus, the manufacture of foods that are safe, wholesome and nutritious are moral and legal obligations. Constant updating of the knowledge of sanitation principles and good manufacturing practices are fundamental requirements. Further, the training of all plant personnel in sanitation principles and the communicating of specifications to all is mandatory.

Sanitation is a way of life and must come from within both management and the individual worker. Sanitation is the moral and legal responsibility of everyone working in a food plant to uphold. Good sanitary principles are the key to production, processing and marketing of safe, wholesome and nutritious foods.

CHAPTER 2

CURRENT GOOD MANUFACTURING PRACTICES - THE REGULATION

Congress in 1938 authorized the Secretary of the Agency to promulgate regulations for the enforcement of the Act. Specifically, in Chapter VII Section 701 (a) "The authority to promulgate regulations for the efficient enforcement of this Act, except as otherwise provided in this section, is hereby vested in the Secretary". This authority was not extensively used prior to the late '60's. In 1969 the Secretary published the first GMP's as Part 128 of the Code of Federal Regulations. In 1977 this was recodified as Part 110 and was revised and updated in 1986.

The CGMP Regulations, Part 110, implement the adulteration provisions of Section 402 of the Federal Food, Drug, and Cosmetic Act. Thus, FDA uses the GMP regulations to control the risk of filth, chemical, microbiological, and other means of contaminating foods during their manufacture. The GMP regulations cover every aspect of food production, employee training, plant design, equipment specifications and cleaning, quality assurance evaluating, and even the distribution of the finished products.

For the readers not familiar with the Code of Federal Regulations, the following may be helpful. The proposed regulations are generally first published in the Federal Register, a daily publication setting forth rules by the Executive Department and the agencies of the Federal Government. Annually, these changes are codified and published as the Code of Federal Regulations. The Code is divided into 50 titles which represent broad areas subject to Federal regulations. Each title is divided into chapters which usually bear the name of the issuing agency.

Each chapter is further subdivided into parts covering specific regulatory areas. Each volume of the Code is revised at least once each calendar year and issued on a quarterly basis approximately as follows:

Title 1 through Title 16..................as of January 1
Title 17 through Title 27..............as of April 1
Title 28 through Title 41..............as of July 1
Title 42 through Title 50...............as of October 1

"To keep up-to-date in changes in the Code of Federal Regulations, one must have access to the Federal Register. To determine whether a Code volume has been amended since its revision date, consult the "List of CFR Sections Affected" which is issued monthly, and the "Cumulative List of Parts Affected," which appears in the Reader Aids section of the daily Federal Register. These two lists will identify the Federal Register page number of the latest amendment of any rule."

Title 7 is for Agriculture with some 36 Chapters. Title 21 is for Food and Drugs with 2 Chapters. The first Chapter has over 1300 Parts and is published in 9 volumes. Volume I contains Part 1-99, Volume II Part 100-169, Volume III Part 170-199, Volume IV Part 200-299, Volume V Part 300-499, Volume VI Part 500-599, Volume VII 600-799, Volume VIII 800-1299 and 1300-end. The first 8 Volumes deal with Food and Drug Administration, Department of Health and Human Services and Volume IX deals with Drug Enforcement Administration, Department of Justice.

Part 110 is found in Volume II or Subchapter B of Chapter I. Subchapter B starts with Part 100 and goes through Part 169, although not all Parts are complete. The reader should be aware of, at least, Part 108 "Emergency Permit Control", Part 109 "Unavoidable contaminants in food for human consumption and food-packaging materials", Part 110 "Current good manufacturing practice in manufacturing, packing, or holding human foods", Part 113 "Thermally processed low-acid foods packaged in hermetically sealed containers", Part 114 "Acidified Foods", and Part 129 "Processing and Bottling of bottled drinking water. Other Parts may be germane, depending on the readers interest.

FIGURE 2.1 - United States Capitol Building Washington, D.C.
Photo courtesy of Washington, D.C. Convention and Vistors Association

Part 110 is reproduced as follows:

PART 110-
CURRENT GOOD MANUFACTURING PRACTICE IN MANUFACTURING, PACKING, OR HOLDING HUMAN FOOD

Promugated as Part 128 4/26/69, effective 5/26/69,
recodified as Part 110 3/15/77.
Revised 51 FR 22458, 6/19/86, Effective 12/16/86.

Title 21 – FOOD AND DRUGS

PART 110 CURRENT FOOD MANFUCTURING PRACTICE (SANITATION) IN MANUFACTURING, PROCESSING, PACKING, OR HOLDING HUMAN FOOD.

Subpart A - General Provisions

Subpart B - Buildings and Facilities

Subpart C - Equipment

Subpart D - Reserved

Subpart E - Production and Process Controls

Subpart F - Reserved

Subpart G - Defect Action Levels

AUTHORITY: Secs. 302, 303, 304, 402(a), 701(a), 52 Stat. 1043-1046 as amended, 1055 (21 U.S.C. 332, 333, 334, 342(a), 371(a)); sec. 361, 58 Stat. 703 (42 U.S.C. 264); 21 CFR 5.10, 5.11.

SOURCE: 51 FR 24475, June 19, 1986, unless otherwise noted.

Subpart A - General Provisions

§ 110.3 Definitions.

The definitions and interpretations of terms in section 201 of the Federal Food, Drug, and Cosmetic Act (the act) are applicable to such terms when used in this part. The following definitions shall also apply:

(a) "Acid foods or acidified foods" means foods that have an equilibrium pH of 4.6 or below.

(b) "Adequate" means that which is needed to accomplish the intended purpose in keeping with good public health practice.

(c) "Batter" means a semifluid substance, usually composed of flour and other ingredients, into which principal components of food are dipped or with which they are coated, or which may be used directly to form bakery foods.

(d) "Blanching," except for tree nuts and peanuts, means a prepackaging heat treatment of foodstuffs for a sufficient time and at a sufficient temperature to partially or completely inactivate the naturally occurring enzymes and to effect other physical or biochemical changes in the food.

(e) "Critical control point" means a point in a food process where there is a high probability that improper control may cause, allow, or contribute to a hazard or to filth in the final food or decomposition of the final food.

(f) "Food" means food as defined in section 201(f) of the act and includes raw materials and ingredients.

(g) "Food-contact surfaces" are those surfaces that contact human food and those surfaces from which drainage onto the food or onto surfaces that contact the food ordinarily occurs during the normal course of operations. "Food-contact surfaces" includes utensils and food-contact surfaces of equipment.

(h) "Lot" means the food produced during a period of time indicated by a specific code.

(i) "Microorganisms" means yeasts, molds, bacteria, and viruses and includes, but is not limited to, species having public health significance. The term "undesirable microorganisms" that are of public health significance, that subject food to decomposition, that indicate that food is contaminated with filth, or that otherwise may cause food to be adulterated within the meaning of the act. Occasionally in these regulations, FDA used the adjective "microbial" instead of using an adjectival phrase containing the word microorganism.

(j) "Pest" refers to any objectionable animals or insects including, but not limited to, birds, rodents, flies, and larvae.

(k) "Plant" means the building or facility or parts thereof, used for or in connection with the manufacturing, packaging, labeling, or holding of human food.

(l) "Quality control operation" means a planned and systematic procedure for taking all actions necessary to prevent food from being adulterated within the meaning of the act.

(m) "Rework" means clean unadulterated food that has been removed from processing for reasons other than insanitary conditions or that has been successfully reconditioned by reprocessing and that is suitable for use as food.

(n) "Safe-moisture level" is a level of moisture low enough to prevent the growth of undesirable microorganisms in the finished product under the intended conditions of manufacturing, storage, and distribution. The maximum safe moisture level for a food is based on its water activity (a_w). An a_w will be considered safe for a food if adequate data are available that demonstrate that the food at or below the given a_w will not support the growth of undesirable microorganisms.

(o) "Sanitize" means to adequately treat food-contact surfaces by a process that is effective in destroying vegetative cells of microorganisms of public health significance, and in substantially reducing numbers of other undesirable microorganisms, but without adversely affecting the product or its safety for the consumer.

(p) "Shall" is used to state mandatory requirements.

(q) "Should" is used to state recommended or advisory procedures or identify recommended equipment.

(r) "Water activity" (a_w) is a measure of the free moisture in a food and is the quotient of the water vapor pressure of the substance divided by the vapor pressure of pure water at the same temperature.

§ 110.5 Current good manufacturing practice.

(a) The criteria and definitions in this part shall apply in determining whether a food is adulterated (1) within the meaning of section 402(a)(3) of the act in that the food has been manufactured under such conditions that it is unfit for food; or (2) within the meaning of section 402(a)(4) of the act in that the food has been prepared, packed, or held under insanitary conditions whereby it may have become contaminated with filth, or whereby it may have been rendered injurious to health. The criteria and definitions in this part also apply in determining whether a food is in violation of section 361 of the Public Health Act (42 U.S.C. 264).

(b) Food covered by specific current good manufacturing practice regulations also is subject to the requirements of those regulations.

§ 110.10 Personnel.

The plant management shall take all reasonable measures and precautions to ensure the following:

(a) *Disease control.* Any person who, by medical examination or supervisory observation, is shown to have, or appears to have, an illness, open lesion, including boils, sores, or infected wounds, or any other abnormal source of microbial contamination by which there is a reasonable possibility of food, food-contact surfaces, or food-packaging materials becoming contaminated, shall be excluded from any operations which may be expected to result in such contamination until the condition is corrected. Personnel shall be instructed to report such health conditions to their supervisors.

(b) *Cleanliness.* All persons working in direct contact with food, food-contact surfaces, and food-packaging materials shall conform to hygienic practices while on duty to the extent necessary to protect against contamination of food. The methods for maintaining cleanliness include, but are not limited to:

(1) Wearing outer garments suitable to the operation in a manner that protects against the contamination of food, food-contact surfaces, or food-packaging materials.

(2) Maintaining adequate personal cleanliness.

(3) Washing hands thoroughly (and sanitizing if necessary to protect against contamination with undesirable microorganisms) in an adequate hand-washing facility before starting work, after each absence from the work station, and at any other time when the hands may have become soiled or contaminated.

(4) Removing all insecure jewelry and other objects that might fall into food, equipment, or containers, and removing hand jewelry that cannot be adequately sanitized during periods in which food is manipulated by hand. If such hand jewelry cannot be removed, it may be covered by material which can be maintained in an intact, clean, and sanitary condition and which effectively protects against the contamination by these objects of the food, food-contact surfaces, or food-packaging materials.

(5) Maintaining gloves, if they are used in food handling, in an intact, clean and sanitary condition. The gloves should be of an impermeable material.

(6) Wearing, where appropriate, in an effective manner, hair nets, headbands, caps, beard covers, or other effective hair restraints.

(7) Storing clothing or other personal belongings in areas other than where food is exposed or where equipment or utensils are washed.

(8) Confining the following to areas other than where food may be exposed or where equipment or utensils are washed: eating food, chewing gum, drinking beverages, or using tobacco.

(9) Taking any other necessary precautions to protect against contamination of food, food-contact surfaces, or food-packaging materials with microorganisms or foreign substances including, but not limited to, perspiration, hair, cosmetics, tobacco, chemicals, and medicines applied to the skin.

(c) *Education and training.* Personnel responsible for identifying sanitation failures or food contamination should have a background of education or experience, or a combination thereof, to provide a level of competency necessary for production of clean and safe food. Food handlers and supervisors should receive appropriate training in proper food handling techniques and food-protection principles and should be informed of the danger of poor personal hygiene and insanitary practices.

(d) *Supervision.* Responsibility for assuring compliance by all personnel with all requirements of this part shall be clearly assigned to competent supervisory personnel.

§ 110.19 Exclusions.

(a) The following operations are not subject to this part: Establishments engaged solely in the harvesting, storage, or distribution of one or more "raw agricultural commodities," as defined in section 201(r) of the act, which are ordinarily cleaned, prepared, treated, or otherwise processed before being marketed to the consuming public.

(b) FDA, however, will issue special regulations if it is necessary to cover these excluded operations.

Subpart B Buildings and Facilities

§ 110.20 Plant and grounds.

(a) *Grounds.* The grounds about a food plant under the control of the operator shall be kept in a condition that will protect against the contamination of food. The methods for adequate maintenance of grounds include, but are not limited to:

(1) Properly storing equipment, removing litter and waste, and cutting weeds or grass within the immediate vicinity of the plant buildings or structures that may constitute an attractive, breeding place, or harborage for pests.

(2) Maintaining roads, yards, and parking lots so that they do not constitute a source of contamination in areas where food is exposed.

(3) Adequately draining areas that may contribute contamination to food by seepage, foot-borne filth, or providing a breeding place for pests.

(4) Operating systems for waste treatment and disposal in an adequate manner so that they do not constitute a source of contamination in areas where food is exposed.

If the plant grounds are bordered by grounds not under the operator's control and not maintained in the manner described in paragraph (a) (1) through (3) of this section, care shall be exercised in the plant by inspection, extermination, or other means to exclude pests, dirt, and filth that may be source of food contamination.

(b) *Plant construction and design.* Plant buildings and structures shall be suitable in size, construction, and design to facilitate maintenance and sanitary operations for food-manufacturing purposes. The plant and facilities shall:

(1) Provide sufficient space for such placement of equipment and storage of materials as is necessary for the maintenance of sani-

tary operations and the production of safe food.

(2) Permit the taking of proper precautions to reduce the potential for contamination of food, food-contact surfaces, or food-packaging materials with microorganisms, chemicals, filth, or other extraneous material. The potential for contamination may be reduced by adequate food safety controls and operating practices or effective design, including the separation of operations in which contamination is likely to occur, by one or more of the following means: location, time, partition, air flow, enclosed systems, or other effective means.

(3) Permit the taking of proper precautions to protect food in outdoor bulk fermentation vessels by any effective means, including:

(i) Using protective coverings.

(ii) Controlling areas over and around the vessels to eliminate harborages for pests.

(iii) Checking on a regular basis for pests and pest infestation.

(iv) Skimming the fermentation vessels, as necessary.

(4) Be constructed in such a manner that floors, walls, and ceilings may be adequately cleaned and kept clean and kept in good repair; that drip or condensate from fixtures, ducts and pipes does not contaminate food, food-contact surfaces, or food-packaging materials; and that aisles or working spaces are provided between equipment and walls and are adequately unobstructed and of adequate width to permit employees to perform their duties and to protect against contaminating food or food-contact surfaces with clothing or personal contact.

(5) Provide adequate lighting in hand-washing areas, dressing and locker rooms, and toilet rooms and in all areas where food is examined, processed, or stored and where equipment or utensils are cleaned; and provide safety-type light bulbs, fixtures, sky-lights, or other glass suspended over exposed food in any step of preparation or otherwise protect against food contamination in case of glass breakage.

(6) Provide adequate ventilation or control equipment to minimize odors and vapors (including steam and noxious fumes)in areas where they may contaminate food; and locate and operate fans and other air-blowing equipment in a manner that minimizes the potential for contaminating food, food packaging materials, and food contact surfaces.

(7) Provide, where necessary, adequate screening or other protection against pests.

110.35 21 CFR Ch. 1 (4-1-88 Edition)

§ 110.35 Sanitary operations.

(a) *General maintenance.* Buildings, fixtures, and other physical facilities of the plant shall be maintained in a sanitary condition and shall be kept in repair sufficient to prevent food from becoming adulterated within the meaning of the act. Cleaning and sanitizing of utensils and equipment shall be conducted in a manner that protects against contamination of food, food-contact surfaces, or food-packaging materials.

(b) *Substances used in cleaning and sanitizing; storage of toxic materials.*

(1) Cleaning compounds and sanitizing agents used in cleaning and sanitizing procedures shall be free from undesirable microorganisms and shall be safe and adequate under the conditions of use. Compliance with this requirement may be verified by any effective means including purchase of these substances under a supplier's guarantee or certification, or examination of these substances for contamination. Only the following toxic materials that are required to maintain sanitary conditions may be used or stored in a plant where food is processed or exposed: (i) Those required to maintain clean and sanitary conditions; (ii) Those necessary for use in laboratory testing procedures; (iii) Those necessary for plant and equipment maintenance and operation; and (iv) Those necessary for use in the plant's operations.

(2) Toxic cleaning compounds, sanitizing agents, and pesticide chemicals shall be identified, held, and stored in a manner that protects against contamination of food, food-contact surfaces, or food-packaging materials. All relevant regulations promulgated by other Federal, State, and local government agencies for the application, use, or holding of these products should be followed.

(c) Pest control. No pests shall be allowed in any area of a food plant. Guard or guide dogs may be allowed in some areas of a plant if the presence of the dogs is unlikely to result in contamination of food, food-contact surfaces, or food-packaging materials. Effective measures shall be taken to exclude pests from the processing areas and to protect against the contamination of food on the premises by pests. The use of insecticides or rodenticide is permitted only under precautions and restrictions that will protect against the contamination of food, food-contact surfaces, and food-packaging materials.

(d) *Sanitation of food-contact surfaces.* All food-contact surfaces including utensils and food-contact surfaces of equipment, shall be cleaned as frequently as necessary to protect against contamination of food.

(1) Food-contact surfaces used for manufacturing or holding low-moisture food shall be in a dry, sanitary condition at the time of use. When the surfaces are wet-cleaned, they shall, when necessary, be sanitized and thoroughly dried before subsequent use.

(2) In wet processing, when cleaning is necessary to protect

against the introduction of microorganisms into food, all food-contact surfaces shall be cleaned and sanitized before use and after any interruption during which the food-contact surfaces may have become contaminated. Where equipment and utensils are used in a continuous production operation, the utensils and food-contact surfaces of the equipment shall be cleaned and sanitized as necessary.

(3) Non-food-contact surfaces of equipment used in the operation of food plants should be cleaned as frequently as necessary to protect against contamination of food.

(4) Single-service articles (such as utensils intended for one-time use, paper cups, and paper towels) should be stored in appropriate containers and shall be handled, dispensed, used, and disposed of in a manner that protects against contamination of food or food-contact surfaces.

(5) Sanitizing agents shall be adequate and safe under conditions of use. Any facility, procedure, or machine is acceptable for cleaning and sanitizing equipment and utensils if it is established that the facility, procedure, or machine will routinely render equipment and utensils clean and provide adequate cleaning and sanitizing treatment.

(e) *Storage and handling of cleaned portable equipment and utensils.* Cleaned and sanitized portable equipment with food-contact surfaces and utensils should be stored in a location and manner that protects food-contact surfaces from contamination.

§ 110.37 Sanitary facilities and controls.

Each plant shall be equipped with adequate sanitary facilities and accommodations including, but not limited to:

(a) *Water supply.* The water supply shall be sufficient for the operations intended and shall be derived from an adequate source. Any water that contacts food or food-contact surfaces shall be safe and of adequate sanitary quality. Running water at a suitable temperature, and under pressure as needed, shall be provided in all areas where required for the processing of food, for the cleaning of equipment, utensils, and food-packaging materials, or for employee sanitary facilities.

(b) *Plumbing.* Plumbing shall be of adequate size and design and adequately installed and maintained to:

(1) Carry sufficient quantities of water to required locations through-out the plant.

(2) Properly convey sewage and liquid disposable waste from the plant.

(3) Avoid constituting a source of contamination to food, water supplies, equipment, or utensils or creating an unsanitary condition.

(4) Provide adequate floor drainage in all areas where floors are subject to flooding-type cleaning or where normal operations release or discharge water or other liquid waste on the floor.

(5) Provide that there is not back-flow from, or cross-connection between, piping systems that discharge waste water or sewage and piping systems that carry water for food or food manufacturing.

(c) *Sewage disposal.* Sewage disposal shall be made into an adequate sewerage system or disposed of through other adequate means.

(d) *Toilet facilities.* Each plant shall provide its employees with adequate, readily accessible toilet facilities. Compliance with this requirement may be accomplished by:

(1) Maintaining the facilities in a sanitary condition.

(2) Keeping the facilities in good repair at all times.

(3) Providing self-closing doors.

(4) Providing doors that do not open into areas where food is exposed to airborne contamination, except where alternate means have been taken to protect against such contamination (such as double doors or positive air-flow systems).

(e) *Hand-washing facilities.* Hand-washing facilities shall be adequate and convenient and be furnished with running water at a suitable temperature. Compliance with this requirement may be accomplished by providing:

(1) Hand-washing and, where appropriate, hand-sanitizing facilities at each location in the plant where good sanitary practices require employees to wash and/or sanitize their hands.

(2) Effective hand-cleaning and sanitizing preparations.

(3) Sanitary towel service or suitable drying devices.

(4) Devices or fixtures, such as water control valves, so designed and constructed to protect against recontamination of clean, sanitized hands.

(5) Readily understandable signs directing employees handling unprotected food, unprotected food-packaging materials, of food-contact surfaces to wash and, where appropriate, sanitize their hands before they start work, after each absence from post of duty, and when their hands may have become soiled or contaminated. These signs may be posted in the processing room(s) and in all other areas where employees may handle such food, materials, or surfaces.

(6) Refuse receptacles that are constructed and maintained in a manner that protects against contamination of food.

(f) *Rubbish and offal disposal.* Rubbish and any offal shall be so conveyed, stored, and disposed of as to minimize the development of odor, minimize the potential for the waste becoming an attractant and harborage or breeding place for pests, and protect against contamination of food, food-contact surfaces, water supplies, and ground surfaces.

Subpart C Equipment

§ 110.40 Equipment and utensils.

(a) All plant equipment and utensils shall be so designed and of such material and workmanship as to be adequately cleanable, and shall be properly maintained. The design, construction, and use of equipment and utensils shall preclude the adulteration of food with lubricants, fuel, metal fragments, contaminated water, or any other contaminants. All equipment should be so installed and maintained as to facilitate the cleaning of the equipment and of all adjacent spaces. Food-contact surfaces shall be corrosion-resistant when in contact with food. They shall be made of nontoxic materials and designed to withstand the environment of their intended use and the action of food, and, if applicable, cleaning compounds and sanitizing agents. Food-contact surfaces shall be maintained to protect food from being contaminated by any source, including unlawful indirect food additives.

(b) Seams on food-contact surfaces shall be smoothly bonded or maintained so as to minimize accumulation of food particles, dirt, and organic matter and thus minimize the opportunity for growth of microorganisms.

(c) Equipment that is in the manufacturing or food-handling area and that does not come into contact with food shall be so constructed that it can be kept in a clean condition.

(d) Holding, conveying, and manufacturing systems, including gravimetric, pneumatic, closed, and automated systems, shall be of a design and construction that enables them to be maintained in an appropriate sanitary condition.

(e) Each freezer and cold storage compartment used to store and hold food capable of supporting growth of microorganisms shall be fitted with an indicating thermometer, temperature-measuring device, or temperature-recording device so installed as to show the temperature accurately within the compartment, and should be fitted with an automatic control for regulating temperature or with an automatic alarm system to indicate a significant temperature change in a manual operation.

(f) Instruments and controls used for measuring, regulating, or recording temperatures, pH, acidity, water activity, or other conditions that control or prevent the growth of undesirable microorganisms in food shall be accurate and adequately maintained, and adequate in number for their designated uses.

(g) Compressed air or other gases mechanically introduced into food or used to clean food-contact surfaces or equipment shall be treated in such a way that food is not contaminated with unlawful indirect food additives.

Subpart D - Reserved

Subpart E - Production and Process Controls

§ 110.80 Processes and controls.

All operations in the receiving, inspecting, transporting, segregating, preparing, manufacturing, packaging, and storing of food shall be conducted in accordance with adequate sanitation principles. Appropriate quality control operations shall be employed to ensure that food is suitable for human consumption and that food-packaging materials are safe and suitable. Overall sanitation of the plant shall be under the supervision of one or more competent individuals assigned responsibility for this function. All reasonable precautions shall be taken to ensure that production procedures do not contribute contamination from any source. Chemical, microbial, or extraneous-material testing procedures shall be used where necessary to identify sanitation failures or possible food contamination. All food that has become contaminated to the extent that it is adulterated within the meaning of the act shall be rejected, or if permissible, treated or processed to eliminate the contamination.

(a) *Raw materials and other ingredients.*

(1) Raw materials and other ingredients shall be inspected and segregated or otherwise handled as necessary to ascertain that they are clean and suitable for processing into food and shall be stored under conditions that will protect against contamination and minimize deterioration. Raw materials shall be washed or cleaned as necessary to remove soil or other contamination. Water used for washing, rinsing, or conveying food shall be safe and of adequate sanitary quality. Water may be reused for washing, rinsing, or conveying food if it does not increase the level of contamination of the food. Containers and carriers of raw materials should be inspected on receipt to ensure that their condition has not contributed to the contamination or deterioration of food.

(2) Raw materials and other ingredients shall either not contain levels of micro-organisms that may produce food poisoning or other disease in humans, or they shall be pasteurized or otherwise treated during manufacturing operations so that they no longer contain levels that would cause the product to be adulterated within the meaning of the act. Compliance with this requirement may be verified by any effective means, including purchasing raw materials and other ingredients under a supplier's guarantee or certification.

(3) Raw materials and other ingredients susceptible to contamination with aflatoxin or other natural toxins shall comply with current Food and Drug Administration regulations, guidelines, and action levels for poisonous or deleterious substances before these materials or ingredients are incorporated into finished food.

Compliance with this requirement may be accomplished by purchasing raw materials and other ingredients under a supplier's guarantee or certification, or may be verified by analyzing these materials and ingredients for aflatoxins and other natural toxins.

(4) Raw materials, other ingredients, and rework susceptible to contamination with pests, undesirable microorganisms, or extraneous material shall comply with applicable Food and Drug Administration regulations, guidelines, and defect action levels for natural or unavoidable defects if a manufacturer wishes to use the materials in manufacturing food. Compliance with this requirement may be verified by any effective means, including purchasing the materials under a supplier's guarantee or certification, or examination of these materials for contamination.

(5) Raw materials, other ingredients, and rework shall be held in bulk, or in containers designed and constructed so as to protect against contamination and shall be held at such temperature and relative humidity and in such a manner as to prevent the food from becoming adulterated within the meaning of the act. Material scheduled for rework shall be identified as such.

(6) Frozen raw materials and other ingredients shall be kept frozen. If thawing is required prior to use, it shall be done in a manner that prevents the raw materials and other ingredients from becoming adulterated within the meaning of the act.

(7) Liquid or dry raw materials and other ingredients received and stored in bulk form shall be held in a manner that protects against contamination.

(b) *Manufacturing operations.*

(1) Equipment and utensils and finished food containers shall be maintained in an acceptable condition through appropriate cleaning and sanitizing, as necessary. Insofar as necessary, equipment shall be taken apart for thorough cleaning.

(2) All food manufacturing, including packaging and storage, shall be conducted under such conditions and controls as are necessary to minimize the potential for the growth of microorganisms, or for the contamination of food. One way to comply with this requirement is careful monitoring of physical factors such as time, temperature, humidity, a_w, pH, pressure, flow rate, and manufacturing operations such as freezing, dehydration, heat processing, acidification, and refrigeration to ensure that mechanical breakdowns, time delays, temperature fluctuations, and other factors do not contribute to the decomposition or contamination of food.

(3) Food that can support the rapid growth of undesirable microorganisms, particularly those of public health significance,

shall be held in a manner that prevents the food from becoming adulterated within the meaning of the act. Compliance with this requirement may be accomplished by any effective means, including:

(i) Maintaining refrigerated foods at 45°F (7.2°C) or below as appropriate for the particular food involved.

(ii) Maintaining frozen foods in a frozen state.

(iii) Maintaining hot foods at 140°F (60°C) or above.

(iv) Heat treating acid or acidified foods to destroy mesophilic microorganisms when those foods are to be held in hermetically sealed containers at ambient temperatures.

(4) Measures such as sterilizing, irradiating, pasteurizing, freezing, refrigerating, controlling pH or controlling a_w that are taken to destroy or prevent the growth of undesirable microorganisms, particularly those of public health significance, shall be adequate under the conditions of manufacture, handling, and distribution to prevent food from being adulterated within the meaning of the act.

(5) Work-in-process shall be handled in a manner that protects against contamination.

(6) Effective measures shall be taken to protect finished food from contamination by raw materials, other ingredients, or refuse. When raw materials, other ingredients, or refuse are unprotected, they shall not be handled simultaneously in a receiving, loading, or shipping area if that handling could result in contaminated food. Food transported by conveyor shall be protected against contamination as necessary.

(7) Equipment, containers, and utensils used to convey, hold, or store raw materials, work-in-process, rework, or food shall be constructed, handled, and maintained during manufacturing or storage in a manner that protects against contamination.

(8) Effective measures shall be taken to protect against the inclusion of metal detectors, or other suitable effective means.

(9) Food, raw materials, and other ingredients that are adulterated within the meaning of the act shall be disposed of in a manner that protects against the contamination of other food. If the adulterated food is capable of being reconditioned, it shall be reconditioned using a method that has been proven to be effective or it shall be reexamined and found not to be adulterated within the meaning of the act before being incorporated into other food.

(10) Mechanical manufacturing steps such as washing, peeling, trimming, cutting, sorting and inspecting, mashing, dewatering, cooling, shredding, extruding, drying, whipping, defatting, and

forming shall be performed so as to protect food against contamination. Compliance with this requirement may be accomplished by providing adequate physical protection of food from contaminants that may drip, drain, or be drawn into the food. Protection may be provided by adequate cleaning and sanitizing of all food-contact surfaces, and by using time and temperature controls at and between each manufacturing step.

(11) Heat blanching, when required in the preparation of food, should be effected by heating the food to the required temperature, holding it at this temperature for the required time, and then either rapidly cooling the food or passing it to subsequent manufacturing without delay. Thermophilic growth and contamination in blanchers should be minimized by the use of adequate operating temperatures and by periodic cleaning. Where the blanched food is washed prior to filling, water used shall be safe and of adequate sanitary quality.

(12) Batters, breading, sauces, gravies, dressings, and other similar preparations shall be treated or maintained in such a manner that they are protected against contamination. Compliance with this requirement may be accomplished by any effective means, including one or more of the following:

(i) Using ingredients free of contamination.

(ii) Employing adequate heat processes where applicable.

(iii) Using adequate time and temperature controls.

(iv) Providing adequate physical protection of components from contaminants that may drip, drain, or be drawn into them.

(v) Cooling to an adequate temperature during manufacturing.

(vi) Disposing of batters at appropriate intervals to protect against the growth of microorganisms.

(13) Filling, assembling, packaging, and other operations shall be performed in such a way that the food is protected against contamination. Compliance with this requirement may be accomplished by any effective means, including:

(i) Use of a quality control operation in which the critical control points are identified and controlled during manufacturing.

(ii) Adequate cleaning and sanitizing of all food-contact surfaces and food containers.

(iii) Using materials for food containers and food-packaging materials that are safe and suitable, as defined in §130.3(d) of this chapter.

(iv) Providing physical protection from contamination, particularly air-borne contamination.

(v) Using sanitary handling procedures.

(14) Food such as, but not limited to, dry mixes, nuts, intermediate moisture food, and dehydrated food, that relies on the control of a_w for preventing the growth of undesirable microorganisms shall be processed to and maintained at a safe moisture level. Compliance with this requirement may be accomplished by any effective means, including employment of one or more of the following practices:

(i) Monitoring the a_w of food.

(ii) Controlling the soluble solids-water ratio in finished food.

(iii) Protecting finished food from moisture pickup, by use of a moisture barrier or by other means, so that the a_w of the food does not increase to an unsafe level.

(15) Food such as, but not limited to, acid and acidified food, that relies principally on the control of pH for preventing the growth of undesirable microorganisms shall be monitored and maintained at a pH of 4.6 or below. Compliance with this requirement may be accomplished by any effective means, including employment of one or more of the following practices:

(i) Monitoring the pH of raw materials, food in process, and finished food.

(ii) Controlling the amount of acid or acidified food added to low-acid food.

(16) When ice is used in contact with food, it shall be made from water that is safe and of adequate sanitary quality, and shall be used only if it has been manufactured in accordance with current good manufacturing practice as outlined in this part.

(17) Food-manufacturing areas and equipment used for manufacturing human food should not be used to manufacture nonhuman food-grade animal feed or inedible products, unless there is no reasonable possibility for the contamination of the human food.

§ 110.93 Warehousing and distribution.

Storage and transportation of finished food shall be under conditions that will protect food against physical, chemical, and microbial contamination as well as against deterioration of the food and the container.

Subpart F - Reserved

Subpart G Defect Action Levels

§ 110.110 Natural or unavoidable defects in food for human use that present no health hazard.

(a) Some foods, even when produced under current good manufacturing practice, contain natural or unavoidable defects that at low levels are not hazardous to health. The Food and Drug Administration establishes maximum levels for these defects in foods produced under current good manufacturing practice and uses these levels in deciding whether to recommend regulatory action.

(b) Defect action levels are established for foods whenever it is necessary and feasible to do so. These levels are subject to change upon the development of new technology or the availability of new information.

(c) Compliance with defect action levels does not excuse violation of the requirement in section 402(a)(4) of the act that food not be prepared, packed, or held under sanitary conditions or the requirements in this part that food manufacturers, distributors, and holders shall observe current good manufacturing practice. Evidence indicating that such a violation exists causes the food to be adulterated within the meaning of the act, even though the amounts of natural or unavoidable defects are lower than the currently established defect action levels. The manufacturer, distributor, and holder of food shall at all times utilize quality control operations that reduce natural or unavoidable defects to the lowest level currently feasible.

(d) The mixing of a food containing defects above the current defect action level with another lot of food is not permitted and renders the final food adulterated within the meaning of the act, regardless of the defect level of the final food.

The Food Defect Action Levels (DAL's) that follow are set on the basis of no hazard to health. Any products that might be harmful to consumers are acted against on the basis of their hazards to health, whether or not they exceed the action levels. In addition, poor manufacturing practices will result in regulatory action, whether the product is above or below the defect action level. The defect action levels are set because it is not possible, and never has been possible, to grow in open fields, harvest, and process crops that are totally free of natural defects. The alternative to establishing natural defect levels in some foods would be to insist on increased utilization of chemical substances to control insects, rodents and other natural contaminants. The alternative is not satisfactory because of the very real danger of exposing consumers to potential hazards from residues of these chemicals, as opposed to the aesthetically unpleasant, but harmless natural and unavoidable defects. The fact that the Food and Drug Administration has an established defect level does not mean that a manufacturer need only stay below that level. The defect action levels do not represent an average of the defects that occur in any of the food categories. The averages are much lower. The levels represent the limit at or above which FDA will take legal action against the product to remove it from the market. The mixing or blending of food containing any amount of defective food at or above the current defect action level with another lot of the same or another food is not permitted and renders the final food unlawful regardless of the defect level of the finished food. The defect action levels on the following list are periodically reviewed and lowered as technology improves. The list is current.

The Method of analysis used by FDA is listed as:

MPM-V ------------ Macroanalytical Procedures Manual
(FDA Technical Bulletin No. 5)

AOAC 44 ----------- Methods of Analysis, 14th Ed., 1984,
Association of Official Analytical Chemists

CHART 2.1 – Food Defect Action Levels

PRODUCT	DEFECT (Method)	ACTION LEVEL
ALLSPICE	Mold (MPM-V32)	Average of 5% or more by weight are moldy
APPLE BUTTER	Mold (AOAC 44.197)	Average of mold count is 12% or more
	Rodent Filth (AOAC 44.086)	Average of 4 or more rodent hairs per 100 grams of apple butter
	Insects (AOAC 44.086)	Average of 5 or more whole or equivalent insects (not counting mites, aphids, thrips, or scale insects) per 100 grams of apple butter
APRICOT, PEACH, AND PEAR NECTARS AND PUREES	Mold (AOAC 44.202)	Average mold count is 12% or more
APRICOTS, CANNED	Insect filth (MPM-V51)	Average of 2% or more by count insect-infested or insect-damaged in a minimum of 10 subsamples.
ASPARAGUS, CANNED OR FROZEN	Insect filth (MPM-V93)	10% by count of spears or pieces are infested with 6 or more attached asparagus beetle eggs and/or sacs
	Insects (MPM-V93)	Asparagus contains an average of 40 or more thrips per 100 grams OR
		Insects (whole or equivalent) of any size average 5 or more per 100 grams OR
		Insects (whole or equivalent) of 3mm or longer have an average aggregate length of 7mm or longer per 100 grams of asparagus
BAY (LAUREL) LEAVES	Mold (MPM-V32)	Average of 5% or more of pieces by weight are moldy
	Insect filth (MPM-V32)	Average of 5% or more pieces by weight are insect-infested
	Mammalian excreta (MPM-V32)	Average of 1 mg or more mammalian excreta per pound after processing
BEETS, CANNED	Rot	Average of 5% or more pieces by weight with dry rot

CHART 2.1 – Food Defect Action Levels - Continued

PRODUCT	DEFECT (Method)	ACTION LEVEL
BERRIES: Drupelet, Canned Frozen (blackberries, raspberries, etc.)	Mold (AOAC 44.205)	Average mold count is 60% or more
	Insects and larvae (AOAC 44.089)	Average of 4 or more larvae per 500 grams OR
		Average of 10 or more whole insects or equivalent per 500 grams (excluding thrips, aphids and mites)
Ligon, Canned	Insect larvae (MPM-V64)	Average of 3 or more larvae per pound in a minimum of 12 subsamples
Multer, Canned	Insects (MPM-V64)	Average of 40 or more thrips per No. 2 can in all subsamples and 20% of subsamples are materially infested
BROCCOLI, FROZEN	Insects and mites (AOAC 44.108)	Average of 60 or more aphids, thrips and/or mites per 100 grams
BRUSSELS SPRTS., FROZEN	Insects (MPM-V95)	Average of 30 or more aphids and/or thrips per 100 grams
CAPSICUM: Pods	Insect filth and/or mold (MPM-V32)	Average of more than 3% of pods by weight are insect-infested and/or moldy
Ground Capsicum (excluding paprika)	Mold (AOAC 44.213)	Average mold count is more than 20%
	Insect filth (AOAC 44.131)	Average of more than 50 insect fragments per 25 grams
	Rodent filth (AOAC 44.131)	Average of more than 6 rodent hairs per 25 grams
Ground Paprika	Mold (AOAC 44.213)	Average mold count is more than 20%
	Insect filth (AOAC 44.146)	Average of more than 75 insect fragments per 25 grams
	Rodent filth (AOAC 44.146)	Average of more than 11 rodent hairs per 25 grams
CASSIA OR CINNAMON BARK (WHOLE)	Mold (MPM-V32)	Average of 5% or more pieces by weight are moldy
	Insect filth (MPM-V32)	Average of 5% or more pieces by weight are insect-infested
	Mammalian excreta (MPM-V32)	Average of more than 1 mg or more mammalian excreta per pound

CHART 2.1 – Food Defect Action Levels - Continued

PRODUCT	DEFECT (Method)	ACTION LEVEL
CHERRIES: Brined and Maraschino	Insect filth (MPM-V48)	Average of 5% or more pieces are rejects due to maggots
Fresh, Canned, or Frozen	Rot (MPM-V48)	Average of 7% or more pieces are rejects due to rot
	Insect filth (MPM-V48)	Average of 4% or more pieces are rejects due to insects other than maggots
CHERRY JAM	Mold (MPM-V61)	Average mold count is 30% or more
CHOCOLATE AND CHOCOLATE LIQUOR	Insect filth (AOAC 44.007)	Average is 60 or more microscopic insect fragments per 100 grams when 6 100-gram subsamples are examined OR
		Any 1 subsample contains 90 or more insect fragments
	Rodent filth (AOAC 44.007)	Average is more than 1 rodent hair per 100 grams in 6 100-gram subsamples examined OR
		Any 1 subsample contains more than 3 rodent hairs
	Shell (AOAC 44.012-13.026)	For chocolate liquor, if the shell is in excess of 2% calculated on the basis of alkali-free nibs
CITRUS FRUIT JUICES, CANNED	Mold (AOAC 44.218)	Average mold count is 10% or more
	Insects and insect eggs (AOAC 44.095 & 44.096)	5 or more Drosophila and other fly eggs per 250 ml or 1 or more maggots per 250 ml
CLOVES	Stems (MPM-V32)	Average of 5% or more stems by weight
COCOA BEANS	Mold (MPM-V18)	More than 4% of beans by count are moldy
	Insect filth (MPM-V18)	More than 4% of beans by count are insect-infested including insect-damaged
	Insect filth and/or mold (MPM-V18)	More than 6% of beans by count are insect-infested or moldy
	Mammalian excreta (MPM-V18)	Mammalian excreta is 10 mg per pound or more

CHART 2.1 – Food Defect Action Levels - Continued

PRODUCT	DEFECT (Method)	ACTION LEVEL
COCOA POWDER PRESS CAKE	Insect filth (AOAC 44.007)	Average of 75 or more microscopic insect fragments per subsample of 50 grams when 6 subsamples are examined OR
		Any 1 subsample contains 125 or more microscopic insect fragments
	Rodent filth (AOAC 44.007)	Average in 6 or more subsamples is more than 2 rodent hairs per subsample of 50 grams OR
		Any 1 subsample contains more than 4 rodent hairs
	Shell (AOAC 44.13.012-13.026)	2% or more shell calculated on the basis of alkali-free nibs
COFFEE BEANS, GREEN	Insect filth and insects (MPM-V1)	Average 10% or more by count are insect-infested or insect-damaged OR
		If live insect infestation is present, 1 live insect in each of 2 or more immediate containers, or 1 dead insect in each of 3 or more immediate containers, or 3 live or dead insects in 1 immediate container AND
		Similar live or dead insect infestation present on or in immediate proximity of the lot OR
		1 or more live insects in each of 3 or more immediate containers OR
		2 or more dead whole insects in 5 or more immediate containers OR
		2 or more live or dead insects on 5 or more of cloth or burlap containers
COFFEE BEANS, GREEN (Con't)	Mold (MPV-V1)	Average of 10% or more beans by count are moldy
COFFEE BEANS, GRADED GREEN	Poor Grade	Beans are poorer than Grade 8 of the New York Green Coffee Association
CONTINENTAL SEEDS OTHER THAN FENNEL SEEDS AND SESAME SEEDS	Mammalian excreta (MPM-V32)	Average of 3 mg or more of mammalian excreta per pound
CORN: SWEET	Insect larvae (corn	2 or more 3mm or longer larvae, cast skins, larval

CHART 2.1 – Food Defect Action Levels - Continued

PRODUCT	DEFECT (Method)	ACTION LEVEL
CORN, CANNED	ear worms, corn borers) (AOAC 44.109)	or cast skin fragments of corn ear worm borer and the aggregate length of such larvae, cast skins, larval or cast skin fragments exceeds 12 mm in 24 pounds (24 No. 303 cans or equivalent)
CORN HUSKS FOR TAMALES	Insect filth (MPM-V115)	Average of 5% or more pieces by weight of the corn husks examined insect-infested (including insect-damaged)
	Mold (MPM-V115)	Average of 5% or more pieces by weight are moldy
CORNMEAL	Insects	Average of 1 or more whole insects (or equivalent) per 50 grams
	Insect filth (AOAC 048)	Average of 25 or more insect fragments per 25 grams
	Rodent filth (AOAC 048)	Average of 1 or more rodent hairs per 25 grams OR
		Average of 1 or more rodent excreta fragment per 50 grams
CRANBERRY SAUCE	Mold (AOAC 44.200)	Average mold count is more than 15% OR
		The mold count of any 1 subsample is more than 50%
CUMIN SEED	Sand and grit (AOAC 44.124)	Average of 9.5% or more ash and/or 1.5% or more acid insoluble ash
CURRENT JAM, BLACK	Mold (MPM-V61)	Average mold count is 75% or more
CURRANTS	Insect filth (MPM-V53)	5% or more by count wormy in the average of the subsamples
CURRY POWDER	Insect filth (AOAC 44.124)	Average of 100 or more insect fragments per 25 grams
	Rodent filth (AOAC 44.124)	Average of 4 or more rodent hairs per 25 grams
DATE MATERIAL (CHOPPED, SLICED, OR MACERATED)	Insects (MPM-V53)	10 or more dead insects in 1 or more subsamples OR
		5 or more dead insects (whole or equivalent) per 100 grams
	Pits (MPM-V53)	2 or more pits and/or pit fragments 2mm or longer measured in the longest dimension per 900 grams

CHART 2.1 – Food Defect Action Levels - Continued

PRODUCT	DEFECT (Method)	ACTION LEVEL
DATES, PITTED	Multiple (MPM-V53)	Average of 5% or more dates by count are rejects (moldy, dead insects, insect excreta, sour, dirty, and/or worthless) as determined by macroscopic suquential examination
	Pits (MPM-V53)	Average of 2 or more pits and/or pit fragments 2 mm or longer in the longest dimension per 100 dates
DATES, WHOLE	Multiple (MPM-V53)	Average of 5% or more dates by count are rejects (moldy, dead insects, insect excreta, sour, dirty, and/or worthless) as determined by macroscopic sequential examination
EGGS AND OTHER EGG PRODUCTS, FROZEN	Decomposition (AOAC 46.003-46.012)	2 or more cans decomposed and at least 2 subsamples from decomposed cans have direct microscopic counts of 5 million or more bacteria per gram
FENNEL SEED	Insects (MPM-V32)	20% or more of subsamples contain insects
	Mammalian excreta (MPM-V32)	20% or more of subsamples contain mammalian excreta OR Average of 3 mg or more of mammalian excreta per pound
FIG PASTE	Insects (AOAC 44.092-44.093)	Over 13 insect heads per 100 grams of fig paste in each of 2 or more subsamples
FIGS	Insect filth and/or mold and/or dirty fruit or pieces of fruit (MPM-V53)	Average of 10% or more pieces by count are rejects Average of more than 10% of pieces are insect-infested and/or moldy dirty fruit or pieces of fruit
FISH, FRESH OR FROZEN (APPLIES TO FISH OR FILLETS WEIGHING 3 POUNDS OR LESS)	Decomposition	Decomposition in 5% or more of the fish or fillets in the sample (but not less than 5) show Class 3 decomposition over at least 25% of their areas OR 20% or more of the fish or fillets in the sample (but not less than 5) show Class 2 decomposition over at least 25% of their areas OR The percentage of fish or fillets showing Class 2 decomposition as above plus 4 times the percentage of those showing Class 3 decomposition as above equals at least 20% and there are at least 5 decomposed fish or fillets in the sample Classes of Decomposition 1. No odor of decomposition 2. Slight odor of decomposition 3. Definite odor of decomposition

CHART 2.1 - Food Defect Action Levels - Continued

PRODUCT	DEFECT (Method)	ACTION LEVEL
FISH, FRESH OR FROZEN - Continued		
Tullibees, Ciscoes, Inconnus, Chubs, and Whitefish	Parasites (cysts) (MPM-V28)	50 parasitic cysts per 100 pounds (whole or fillets), provided that 20% of the fish examined are infested
Blue Fin and other Fresh Water Herring	Parasites (cysts) (MPM-V28)	60 cysts per 100 fish (fish averaging 1 pound or less) or 100 pounds of fish (fish averaging over 1 pound), provided that 20% of the fish examined are infested
Red Fish and Ocean Perch	Parasites (copepods) (MPM-V28)	3% of the fillets examined contain 1 or more copepods accompanied by pus pockets
GINGER, WHOLE	Insect filth and/or mold (MPM-V32)	Average of 3% or more pieces by weight are insect-infested and/or moldy
	Mammalian excreta (MPM-V32)	Average of 3 mg or more of mammalian excreta per pound
GREENS, CANNED	Mildew	Average of 10% or more of leaves, by count or weight, showing mildew over 1/2" in diameter
HOPS	Insects (aphids) (AOAC 44.009)	Average of more than 2,500 aphids per 10 grams
MACARONI AND NOODLE PRODUCTS	Insect filth (AOAC 44.069)	Average of insect fragments equals or exceeds 225 per 225 grams in 6 or more subsamples
	Rodent filth (AOAC 44.069)	Average of rodent hairs equals or exceeds 4.5 per 225 grams in 6 or more subsamples
MACE	Insect filth and/or mold (MPM-V32)	Average of 3% or more pieces by weight are insect-infested and/or moldy
	Mammalian excreta (MPM-V32)	Average of 3 mg or more of mammalian excreta per pound
	Foreign matter (MPM-V32)	Average of 1.5% or more of foreign matter through a 20-mesh sieve
MUSHROOMS, CANNED AND DRIED	Insects (AOAC 44.115 & 44.116)	Average of 20 or more maggots of any size per 100 grams of drained mushrooms and proprotionate liquid or 15 grams of dried mushrooms OR
		Average of 5 or more maggots 2mm or longer per 100 grams of drained mushrooms and proportionate liquid or 15 grams of dried mushrooms

CHART 2.1 – Food Defect Action Levels - Continued

PRODUCT	DEFECT (Method)	ACTION LEVEL
MUSHROOMS, CANNED AND DRIED - Continued	Mites (AOAC 44.115 & 44.116)	Average of 75 mites per 100 grams drained mushrooms and proportionate liquid or 15 grams of dried mushrooms
	Decomposition (MPM-V100)	Average of more than 10%, by weight, of mushrooms are decomposed
NUTMEG	Insect filth and/or mold (MPM-V41)	Average of 10% or more pieces by count are insect-infested and/or moldy
NUTS, TREE	Multiple defects (MPM-V81)	Reject nuts (insect-infested, rancid, moldy, gummy, and shriveled or empty shells) as determined by macroscopic examination at or in excess of the following levels

	Unshelled %	Shelled %
Almonds	5	5
Brazils	10	5
Cashew	-	5
Green Chestnuts	5	-
Baked Chestnuts	10	-
Dried Chestnuts	-	5
Filberts	10	5
Lichee Nuts	15	-
Pecans	10	5
Pili Nuts	15	10
Pistachios	10	5
Walnuts	10	5

PRODUCT	DEFECT (Method)	ACTION LEVEL
OLIVES: Pitted	Pits (MPM-V67)	Average of 1.3 or more by count of olives with whole pits and/or pit fragments of 2 mm or longer measured in the longest dimension
Imported Green	Insect damage (MPM-V67)	7% or more by count showing damage by olive fruit fly
Salad	Pits (MPM-V67)	Average of 1.3 or more by count of olives with whole pits and/or pit fragments 2 mm or longer measured in the longest dimension
	Insect damage (MPM-V67)	9% or more by count showing damage by olive fruit fly
Salt-cured	Insects (MPM-V71)	Average of 6 subsamples is 10% or more olives by count with 10 or more scale insects each
	Mold (MPM-V67)	Average of 6 subsamples is 25% or more olives by count are moldy

CHART 2.1 – Food Defect Action Levels - Continued

PRODUCT	DEFECT (Method)	ACTION LEVEL
OLIVES - Continued		
Imported Black	Insect damage (MPM-V67)	10% or more by count showing damage by olive fruit fly
PEACHES, CANNED AND FROZEN	Moldy or Wormy (MPM-V51)	Average of 3% or more fruit by count are wormy or moldy
	Insect damage (MPM-V51)	In 12 1-pound cans or equivalent, one or more larvae and/or larval fragments whose aggregate length exceeds 5 mm
PEANUT BUTTER	Insect filth (AOAC 44.037)	Average of 30 or more insect fragments per 100 grams
	Rodent filth (AOAC 44.0354)	Average of 1 or more rodent hairs per 100 grams
	Grit (AOAC 44.034)	Gritty taste and water insoluble inorganic residue is more than 25 mg per 100 grams
PEANUTS, SHELLED	Multiple defects (MPM-V89)	Average of 5% or more kernels by count are rejects (insect-infested, moldy, rancid, otherwise decomposed, blanks, and shriveled)
	Insects (MPM-V89)	Average of 20 or more whole insects or equivalent in 100-pound bag siftings
PEANUTS, UNSHELLED	Multiple defects (MPM-V89)	Average of 10% or more peanuts by count are rejects (insect-infested, moldy, rancid, otherwise decomposed, blanks, and shriveled)
PEAS: BLACK-EYED, COWPEAS, FIELDPEAS, DRIED	Insect damage (MPM-V104)	Average of 10% or more by count of class 6 damage or higher minimum of 12 subsamples
PEAS, COWPEAS, BLACK-EYED PEAS (SUCCULENT), CANNED	Insect larvae (MPM-V104)	Average of 5 or more cowpea curculio larvae or the equivalent per No. 2 can
PEAS AND BEANS, DRIED	Insect filth (MPM-V104)	Average of 5% or more by count insect-infested and/or insect-damaged by storage insects in minimum of 12 subsamples
PEPPER, WHOLE	Insect filth and/or mold (MPM-V39)	Average of 1% or more pieces by weight are insect-infested and/or moldy
	Mammalian excreta (MPM-V39)	Average of 1% or more mammalian excreta per pound
	Foreign matter (MPM-V39)	Average of 1% or more pickings and siftings by weight

CHART 2.1 - Food Defect Action Levels - Continued

PRODUCT	DEFECT (Method)	ACTION LEVEL
PINEAPPLE, CANNED	Mold (AOAC 44.199)	Average mold count for 6 subsamples is 20% or more OR
		The mold count of any 1 subsample is 60% or more
PINEAPPLE JUICE	Mold (AOAC 44.199)	Average mold count for 6 subsamples is 15% or more OR
		The mold count of any 1 subsample is 40% or more
PLUMS, CANNED	Rot (MPM-V51)	Average of 5% or more plums by count with rot spots larger than the area of a circle 12 mm in diameter
POPCORN	Rodent filth (AOAC 44.099)	1 or more rodent excreta pellets are found in 1 or more subsamples, and 1 or more rodent hairs are found in 2 or more other subsamples OR
		2 or more rodent hairs per pound and rodent hair is found in 50% or more of the subsamples OR
		20 or more gnawed grains per pound and rodent hair is found in 50% or more of the subsamples
	Field corn	5% or more by weight of field corn
POTATO CHIPS	Rot (MPM-V113)	Average of 6% or more pieces by weight contain rot
PRUNES, DRIED AND DEHYDRATED, LOW-MOISTURE	Multiple defects (MPM-V53)	Average of a minimum of 10 subsamples is 5% or more prunes by count are rejects (insect-infested, moldy or decomposed, dirty, and/or otherwise unfit)
PRUNES, PITTED	Pits (MPM-V53)	Average of 10 subsamples is 2% or more by count with whole pits and/or fragments 2 mm or longer and 4 or more of 10 subsamples of pitted prunes have 2% or more by count with whole pits and/or pit fragments 2 mm or longer
RAISINS	Mold (MPM-V76)	Average of 10 subsamples is 5% or more raisins by count are moldy
	Sand and Grit (MPM-V76)	Average of 10 subsamples is 40 mg or more of sand and grit per 100 grams of natural or golden bleached raisins
	Insects and Insect eggs (AOAC 44.097 & MPM-V76)	10 or more whole or equivalent insects and 35 Drosophila eggs per 8 oz. of golden bleached raisins

CHART 2.1 – Food Defect Action Levels - Continued

PRODUCT	DEFECT (Method)	ACTION LEVEL
SALMON, CANNED	Decomposition	2 or more Class 3 defective cans, regardless of lot or container size OR
		2 to 30 Class 2 and/or Class 3 defective cans are required by sampling plan based on lot size and container size
		(A defective can is defined as one that contains Class 2 or Class 3 decomposition--see FISH product listing.) Sampling plan tables are available on request from FDA)
SESAME SEEDS	Insect filth (MPM-V32)	Average of 5% or more seeds by weight are insect-infested
	Mold (MPM-V32)	Average of 5% or more seeds by weight are decomposed
	Mammalian excreta (MPM-V32)	Average of 5 mg or more mammalian excreta per pound
	Foreign matter (MPM-V32)	Average of 0.5% or more foreign matter by weight
SHRIMP: FRESH OR FROZEN, RAW, HEADLESS PEELED OR BREADED	Decomposition	5% or more are Class 3 or 20% or more are Class 2 decomposition as determined by organoleptic examination OR
		If percentage of Class 2 shrimp plus 4 times percent of Class 3 equals or exceeds 20% (See FISH product listing for definition of decomposition classes)
SHRIMP: IMPORTED CANNED OR COOKED/FROZEN	Decomposition	Indole levels in two or more subsamples equal or exceed 25 micrograms per 100 grams for both original and check analysis
SPICES, LEAFY, OTHER THAN BAY LEAVES	Insect filth and/or mold (MPM-V32)	Average of 5% or more pieces by weight are insect-infested and/or moldy
	Mammalian excreta (MPM-V32)	Average of 1 mg or more of Mammalian excreta per pound after processing
SPINACH, CANNED OR FROZEN	Insects and mites (AOAC 44.110)	Average of 50 or more aphids and/or thrips and/or mites per 100 grams OR
		2 or more 3 mm or longer larvae and/or larval fragments of spinach worms (caterpillars) whose aggregate length exceeds 12 mm are present in 24 pounds OR
		Leaf miners of any size average 8 or more per 100 grams or leaf miners 3 mm or longer average 4 or more per 100 grams

CHART 2.1 – Food Defect Action Levels - Continued

PRODUCT	DEFECT (Method)	ACTION LEVEL
STRAWBERRIES: FROZEN WHOLE OR SLICED	Mold (AOAC 44.205)	Average mold count of 45% or more and mold count of at least half of the subsamples is 55% or more
	Grit	Berries taste gritty
TOMATOES, CANNED	Drosophila fly (AOAC 44.119)	Average of 10 or more fly eggs per 500 grams; or 5 or more fly eggs and 1 or more maggots per 500 grams; or 2 or more maggots per 500 grams, in a minimum of 12 subsamples
TOMATOES, CANNED, WITH OR WITHOUT JUICE (BASED ON DRAINED JUICE)	Mold (AOAC 44.206)	Average mold count in 6 subsamples is more than 15% and the mold counts of all of the subsamples are more than 12%
TOMATOES, CANNED, PACKED IN TOMATO PUREE (BASED ON DRAINED LIQUID)	Mold (AOAC 44.206)	Average mold count in 6 subsamples is more than 29% and the counts of all of the subsamples are are more than 25%
TOMATO JUICE	Drosophila fly (AOAC 44.119)	Average of 10 or more fly eggs per 100 grams; or 5 or more fly eggs and 1 or more maggots per 100 grams; or 2 or more maggots per 100 grams, in a minimum of 12 subsamples
	Mold (AOAC 44.207)	Average mold count in 6 subsamples is 24% or more and the mold counts of all of the subsamples are more than 20%
TOMATO PASTE, PIZZA AND OTHER SAUCES	Drosophila fly (AOAC 44.119)	Average of 30 or more fly eggs per 100 grams; or 15 or more fly eggs and 1 or more maggots per 100 grams; or 2 or more maggots per 100 grams, in a minimum of 12 subsamples
TOMATO PUREE	Drosophila fly (AOAC 44.119)	Average of 20 or more fly eggs per 100 grams; or 10 or more fly eggs and 1 or more maggots per 100 grams; or 2 or more maggots per 100 grams, in a minimum of 12 subsamples
TOMATO PASTE OR PUREE	Mold (AOAC 44.207)	Average mold count in 6 subsamples is 45% or more and the mold counts of all of the subsamples are more than 40%
PIZZA AND OTHER SAUCES	Mold (AOAC 44.209)	Average mold count in 6 subsamples is more than 34% and the counts of all of the subsamples are more than 30%
TOMATO SAUCE (UNDILUTED)	Mold (AOAC 44.207)	Average mold count in 6 subsamples is 45% or more and the mold counts of all of the subsamples are more than 40%

CHART 2.1 - Food Defect Action Levels - Continued

PRODUCT	DEFECT (Method)	ACTION LEVEL
TOMATO CATSUP	Mold (AOAC 44.207)	Average mold count in 6 subsamples is 55% or more
TOMATO POWDER (EXCEPT SPRAY-DRIED)	Mold (AOAC 44.211)	Average mold count in 6 subsamples is 45% or more and the mold counts of all of the subsamples are more than 40%
TOMATO POWDER SPRAY-DRIED	Mold (AOAC 44.211)	Average mold count in 6 subsamples is 67% or more
TOMATO SOUP AND TOMATO PRODUCTS	Mold (AOAC 44.208)	Average mold count in 6 subsamples is 45% or more and the mold counts of all of the subsamples are more than 40%
TUNA, CANNED: ALBACORE, SKIPJACK, AND YELLOWFIN	Histamine (AOAC 18.067)	Histamine content per subsample equals or exceeds 20 mg per 100 grams in both original and check analysis (two subsamples minimum)
WHEAT	Insect damage (MPM-V15)	Average of 32 or more insect-damaged kernels per 100 grams
	Rodent filth (MPM-V15)	Average of 9 mg or more rodent excreta pellets and/or pellet fragments per kilogram
WHEAT FLOUR	Insect filth (AOAC 44.052)	Average of 75 or more insect fragments per 50 grams in 6 subsamples
	Rodent filth (AOAC 44.052)	Average of 1 or more rodent hairs per 50 grams in 6 subsamples

CHAPTER 3

TYPES OF FOOD PLANT INSPECTIONS

The types of food plant inspections can be broadly categorized as follows:

A. Firm Inspection
    1. Quality Assurance/Sanitarian
    2. Management Committee
    3. Corporate Staff
B. Sanitation Supplier Inspection
C. Outside Consultant Inspection
D. Official Inspection
    1. Municipal
    2. State
    3. Food and Drug Administration (FDA)
        a-Routine Establishment Inspection
        b-Directed Inspection
        c-Complaint/Referral Inspection
        d-Product Recall Inspection
        e-CGMP Inspection
    4. U. S. Department of Agriculture (USDA)
        a-Voluntary Inspection
        b-Continuous Inspection
            1-Voluntary Inspection
            2-Compulsory Inspection

A firm's self inspection by production management, Quality Assurance personnel and/or the Sanitarian should be a daily routine inspection of all incoming materials, ingredients, products in process and/or in the warehouse, equipment, personnel, the facilities, and the shipping vehicles. This type of inspection should be carefully performed with check sheets for Satisfactory/Needs Improvement/Unsatisfactory and the reasons why it needs improvement or that it is unsatisfactory (see Figure 3.1) The operator/Quality Assurance personnel/Sanitarian should have the authority to shut down the line or the plant if conditions are not corrected within a given time period. The QA Manager/Sanitarian should be firm and they should provide know-how to the cleaning personnel, supervisors, and the line operators. Sanitation is everyone's business in a food plant and the inspectors are only there to audit the efforts of all concerned. They can be the most useful and helpful people in the entire plant to keep a firm in compliance with the sanitation policy of a firm and out of trouble with the consumer and regulatory personnel during the production of safe and wholesome food.

Figure 3.1 – The Sanitation Committee must identify the potential hazards in the production lines.

A Management Committee Inspection of a given firm should be routinely completed at least once every month. The above same areas should be covered. A more extensive record form (See Figure 3.2) should be used. The committee should be made up of the following firm's personnel:

1. Sanitation Manager
2. Quality Assurance Manager
3. Production Manager
4. Plant Engineer
5. Human Resources/Personnel Manager
6. Safety Manager

In addition, the Union Representative may be added if a plant is unionized and the Safety Manager if different from the personnel manager. These persons must follow the same inspection practices, but in more depth than was used in the QA/Sanitarian inspection. The committee should write a monthly report giving specific time intervals for correction of any Unsatisfactory or Needs Improvement situation. The Committee reports directly to Top Management and if the committee is used properly it can be most valuable to a firm in keeping it in total compliance.

The technical Sanitation Supplier is generally a most knowledgeable individual in the area of sanitation. He or she could be biased because of the business they are in; however, their bias can work in favor of the firm in most cases. The Sanitation Supplier is generally highly trained in all aspects of sanitation and the firm's sanitarian and Quality Assurance Manager can gain valuable information by working with this individual. The Sanitation Supplier has the advantage of being in many types of food plants and he or she can positively make sound recommendations to a firm to keep them in the business of manufacturing safe and wholesome foods. Generally, the Sanitation Supplier is one of the best sources of help available to the food industry. They should be utilized at least once per year to make a plant inspection with or without the Sanitarian/Quality Assurance Manager assistance.

The Outside Consultant is similar to the Sanitation Supplier except he or she generally has no supplies to sell the firm. They are like a "a third party" inspection and the use of the Outside Consultant is most advantageous as in-house personnel may not always see the "trees for the forest". The Outside Consultant is more thorough than most in-house inspectors and he or she can be of great assistance in pin-pointing potential problem areas before they do create a problem. The Outside Consultant should be brought in on a routine random basis, at least, once per year. The Outside Consultant can be of valuable assistance to a firm

FIGURE 3.2 – The inspector identifies the major areas for compliance

and they can follow through to help a firm if difficulty with sanitation problems should occur. The Outside Consultant should review (audit) the Quality Assurance/Sanitation Daily Report and make appropriate recommendations where needed. Further, he or she should review the Committee's Monthly Report and likewise make appropriate recommendations where needed. The value of the Outside Consultant to a firm is only as good as the firm uses these recommendations.

Official Inspections are authorized by Municipalities, States and, of course, the Federal Government. The Food, Drug and Cosmetic Act of 1938 gives the authority to the agency in Chapter VII, as follows:

SEC. 704. (a)(1) For purposes of enforcement of this Act, officers or employees duly designated by the Secretary, upon presenting appropriate credentials and a written notice to the owner, opera-

tor, or agent in charge, are authorized (A) to enter, at reasonable times, any factory, warehouse, or establishment in which food, drugs, devices, or cosmetics are manufactured, processed, packed or held, for introduction into interstate commerce or are held after such introduction, or to enter any vehicle being used to transport or hold such food, drugs, devices, or cosmetics in interstate commerce; and (B) to inspect, at reasonable times and within reasonable limits and in a reasonable manner, such factory, warehouse, establishment, or vehicle and all pertinent equipment, finished and unfinished materials, containers, and labeling therein. A separate notice shall be given for each such inspection, but a notice shall not be required for each entry made during the period covered by the inspection. Each such inspection shall be commenced and completed with reasonable promptness.

(b) Upon completing of any such inspection of a factory, warehouse, or other establishment, and prior to leaving the premises, the officer or employee making the inspection shall give to the owner, operator, or agent in charge a report in writing setting forth any conditions or practices observed by him which, in his judgment, indicate that any food, drug, device, or cosmetic in such establishment (1) consists in whole or in part of any filthy, putrid, or decomposed substance, or (2) has been prepared, packed, or held under insanitary conditions whereby it may have become contaminated with filth, or whereby it may have been rendered injurious to health. A copy of such report shall be sent promptly to the Secretary

(c) If the officer or employee making any such inspection of a factory, warehouse, or other establishment has obtained any sample in the course of the inspection, upon completion of the inspection and prior to leaving the premises he shall give to the owner, operator, or agent in charge a receipt describing the samples obtained.

(d) Whenever in the course of any such inspection of a factory or other establishment where food is manufactured, processed, or packed, the officer or employee making the inspection obtains a sample of any such food, and an analysis is made of such sample for the purpose of ascertaining whether such food consists in whole or in part of any filthy, putrid, or decomposed substance, or is otherwise unfit for food, a copy of the results of such analy-

sis shall be furnished promptly to the owner, operator, or agent in charge.

The FDA routine establishment inspection is the basic tool to determine if goods are in compliance with the law and to obtain evidence to support legal action wherein violations are found. The extent of this type of inspection depends on current compliance programs of FDA, problems within the industry, and conditions found as the inspection progresses. This type of inspection may be an "Abbreviated Surveillance Inspection" (ASI) looking at critical factors as opposed to the Hazardous Analysis Critical Control Point (HACCP) inspection wherein critical control points are used to identify dangers to public health. (See Chapter 22). The Abbreviated Surveillance Inspection usually results in an abbreviated report wherein any minor objectional condition are noted on the "I Observed" report (Form FD-483) (See Figure 3.3) If any conditions are noted on the above report, they shall be brought into compliance within a specified time.

Directed Inspections are the result of consumer complaints, product recalls, or other events that may cause FDA to inspect a facility. Such Directed Inspections concentrate on one or more critical points of a plant's operation and are designed to determine the specific cause of a suspected problem or violation.

Consumer Complaint/Referral Inspections occur when FDA receives a consumer complaint against a firm or product, or another federal or state agency refers a complaint to FDA. Immediate corrective action is required by the firm.

Contaminated Product Recall Inspections are conducted when FDA has requested that a manufacturer or distributor recall a product that has violated the FD&C Act. Such inspections are conducted to assure that the violative product has been removed from the marketplace and kept out of distribution. If the products are still being distributed, FDA can initiate seizure or injunction proceedings.

Current Good Manufacturing Practices Inspections (CGMP's Inspections) are conducted by FDA to ensure that the company and responsible individuals are aware of and adhering to FDA developed CGMP's.

The USDA Inspections may be Voluntary or Continuous. Voluntary Inspections are commonly found with fruit and vegetable processing firms and may involve the complete operation or just the finished products. Continuous inspections, also, may be voluntary, particularly for fruits and vegetables, or they may be compulsory as they are for meats, poultry and eggs. Continuous Inspections generally involve all the conditions previously discussed, that is, all raw materials, ingredients, packages, facilities, equipment, personnel and warehousing and shipping. Continuous inspection generally permit the use of the USDA insignia if the firm and products are in compliance.

Many municipalities and most of the States follow the above type inspections as discussed under Federal programs, but generally not as extensive, and of course, they are concerned only within their municipalities or state boundaries.

Regardless of the type of inspection, a firm must make inspections to audit what goes on in a food plant, to know the quality and condition of the raw materials, to establish the safety of the packaging materials and the processing methods, to assure the hygienic nature of the personnel, to evaluate the condition of the equipment and facilities, and generally to assure the performance of operations in the production of safe and wholesome foods. Inspection is like insurance, you should always have it, but hope you never have to use it.

Personnel doing inspections must be trained in quality assurance, sanitation principles, chemistry of cleaning, types and kinds of food plant microorganisms, and the art of communications. It is perhaps the most demanding position in a food plant and it requires persons that are dedicated, sincere, thorough, and cooperative. The example the inspector(s) set is most important to lead all personnel in the need for and the requirements of a clean plant producing clean, safe and wholesome foods.

FORM 3.1 – Daily Sanitation Record Sheet
For Food Processing Plants

Rating Symbols S = Satisfactory Date: ________ Time: ________
NI = Needs Improvement
U = Unsatisfactory Inspector: ________________

AREA AND UNIT	OPERATION	S	NI	U	EXPLANATION
PREMISES:	1. Outside Areas				
	2. Parking lot				
	3. Waste Disposal				
RECIEVING:	1. Containers				
	2. Dumpers & Conveyors				
	3. Floors, Gutters, Walls				
	4. Storages				
	5. Materials				
	6. Docks				
PREPARATION:	1. Flumes				
	2. Belts, Elevators, Conveyors				
	3. Washers				
	4. Peelers				
	5. Slicers				
	6. Floors, Gutters, Walls				
PROCESSING:	1. Conveyors				
	2. Tanks and Pipes				
	3. Fryers				
	4. Floors, Gutters, Walls				
	5. Hoods				
PACKAGING:	1. Conveyors				
	2. Fillers				
	3. Containers				
	4. Packaging Machines				
WAREHOUSING:	1. Pallets				
	2. Floors				
	3. Walls				
	4. Docks				
	5. Trucks				
REST ROOMS:	1. Wash Basins				
	2. Toilets and Urinals				
	3. Supplies				
	4. Floors, Walls				
	5. Lockers				
PERSONAL:	1. Cleanliness				
	2. Head Covering				
	3. Smoking, Food, Etc.				
SHOP ROOM:					
SALES OFFICE:					

FORM 3.2 – Monthly Sanitation Inspection Report
In A Potato Chip Plant

S = Satisfactory NI = Needs Improvement U = Unsatisfactory

Date: ______________

A. Processing Area/Time				
I. a. Potato storage				
b. Floors				
c. Walls, ceilings and lights				
d. Equipment				
e. Other				
f. Remarks				
II. a. Preparation room				
b. Floors				
c. Walls, ceiling and lights				
d. Sewers				
e. Peelers				
f. Slicers				
g. Conveyors				
h. Washers				
i. Other				
j. Remarks				

Form Continued On Next Page

FORM 3.2 – Monthly Sanitation Inspection Report
In A Potato Chip Plant - Continued

S = Satisfactory NI = Needs Improvement U = Unsatisfactory

III. a. Cooking room				
b. Floors				
c. Walls, ceiling and lights				
d. Sewers				
e. Potato chip fryers				
f. Salters				
g. Can washers				
h. Garabage disposal				
i. Conveyors				
IV. a. Boiler room				
b. Floors				
c. Walls, ceilings & lights				
d. Equipment				
e. Painted surfaces				
V. a. Laboratory				
b. Floors				
c. Walls, ceiling and lights				
d. Equipment				
e. Cleanliness				
f. Housekeeping				
VI. Rear Dock				
VII. Shop Area				

Form Continued On Next Page

FORM 3.2 – Monthly Sanitation Inspection Report
In A Potato Chip Plant - Continued

S = Satisfactory NI = Needs Improvement U = Unsatisfactory

B. Packing Area				
I. a. Packaging room				
b. Foors				
c. Walls, ceilings and lights				
d. Sewers				
e. Packaging machine				
line 1				
line 2				
line 3, etc.				
f. Overhead chip conveyor				
g. Plant cans				
h. Customer's cans				
i. Packaging materials				
II. a. Storage room				
b. Floors				
c. Walls, ceiling and lights				
d. Materials stored and stacked properly (bags, roll stock, etc.)				
III. Men's locker room				
IV. Women's locker room				
V. a. Lunch Room				
b. Floors				
c. Tables				

Form Continued On Next Page

FORM 3.2 - Monthly Sanitation Inspection Report
In A Potato Chip Plant - Continued

S = Satisfactory NI = Needs Improvement U = Unsatisfactory

V. Lunch Room - Continued				
d. Chairs				
e. Other				
C. Warehouse Department				
I. a. Warehouse areas				
b. Floors				
c. Walls, ceiling and lights				
d. Empty carton storage				
e. Orderly arrangement of chip stock				
f. Orderly arrangement of miscellaneous stock				
g. Housekeeping				
II. a. Dock area				
b. Refuse cans				
c. Refuse Dumpusters				
D. Premises				
a. Entrance				
b. Parking lots				
c. Offices				
d. Buildings				
e. Others				

Form Continued On Next Page

FORM 3.2 - Monthly Sanitation Inspection Report
In A Potato Chip Plant - Continued

S = Satisfactory NI = Needs Improvement U = Unsatisfactory

E. Personal				
a. Cleanliness				
b. Head coverage				
c. Smoking				
d. Food				
e. Other				

FORM 3.3 – FDA Food Plant Inspection Form - Front

DEPARTMENT OF HEALTH AND HUMAN SERVICES PUBLIC HEALTH SERVICE FOOD AND DRUG ADMINISTRATION	1. DISTRICT ADDRESS & PHONE NO.

TO		
	2. NAME AND TITLE OF INDIVIDUAL	3. DATE
	4. FIRM NAME	5. HOUR a.m.
	6. NUMBER AND STREET	p.m.
	7. CITY AND STATE & ZIP CODE	8. PHONE # & AREA CODE

Notice of Inspection is hereby given pursuant to Section 704(a)(1) of the Federal Food, Drug, and Cosmetic Act [21 U.S.C. 374(a)][1] and/or Part F or G, Title III of the Public Health Service Act [42 U.S.C. 262-264] [2]

9. SIGNATURE *(Food and Drug Administration Employee(s))*	10. TYPE OR PRINT NAME AND TITLE *(FDA Employee(s))*

Applicable portions of Section 704 and other Sections of the Federal Food, Drug, and Cosmetic Act [21 U.S.C. 374] are quoted below:

[1]**Sec. 704. (a)(1) For purposes of enforcement of this Chapter, officers or employees duly designated by the Secretary, upon presenting approp**riate credentials and a written notice to the owner, operator, or agent in charge, are authorized (A) to enter, at reasonable times, any factory, warehouse, or establishment in which food, drugs, devices, or cosmetics are manufactured, processed, packed, or held, for introduction into interstate commerce or after such introduction, or to enter any vehicle being used to transport or hold such food, drugs, devices, or cosmetics in interstate commerce; and (B) to inspect, at reasonable times and within reasonable limits and in a reasonable manner, such factory, warehouse, establishment, or vehicle and all pertinent equipment, finished and unfinished materials, containers, and labeling therein. In the case **of any factory, warehouse, establishment, or consulting laboratory in which prescription drugs or restricted devices are manufactured, processed, packed, or held, the inspection shall extend to all things therein** *(including records, files, papers, processes, controls, and facilities)* **bearing on whether prescription drugs or restricted devices which are adulterated or misbranded within the meaning of this Chapter, or which may not be manufactured, introduced into interstate commerce, or sold, or offered for sale by reason of any provision of this Chapter, have been or are being manufactured, processed, packed, transported, or held in any such place, or otherwise bearing on violation of this Chapter. No inspection authorized by the preceding sentence or by paragraph (3) shall extend to financial data, sales data other than shipment data, pricing data, personnel data** *(other than data as to qualifications of technical and professional personnel performing functions subject to this Act)*, **and research data** *(other than data, relating to new drugs, antibiotic drugs and devices and, subject to reporting and inspection under regulations lawfully issued pursuant to section 505(i) or (k), section 507(d) or (g), section 519, or 520(g), and data relating to other drugs or devices which in the case of a new drug would be subject to reporting or inspection under lawful regulations issued pursuant to section 505(k) of the title.* **A separate notice shall be given for each such inspection, but a notice shall not be required for each entry made during the period covered by the inspection. Each such inspection shall be commenced and completed with reasonable promptness.**

Sec. 704(e) Every person required under section 519 or 520(g) to maintain records and every person who is in charge or custody of such records shall, upon request of an officer or employee designated by the Secretary, permit such officer or employee at all reasonable times to have access to and to copy and verify, such records.

Section 512 (I)(1) In the case of any new animal drug for which an approval of an application filed pursuant to subsection (b) is in effect, the applicant shall establish and maintain such records, and make such reports to the Secretary, of data relating to experience and other data or information, received or otherwise obtained by such applicant with respect to such drug, or with respect to animal feeds bearing or containing such drug, as the Secretary may by general regulation, or by order with respect to such application, prescribe on the basis of a finding that such records and reports are necessary in order to enable the Secretary to determine, or facilitate a determination, whether there is or may be ground for invoking subsection (e) or subsection (m)(4) of this section. Such regulation or order shall provide, where the Secretary deems it to be appropriate, for the examination, upon request, by the persons to whom such regulation or order is applicable, of similar information received or otherwise obtained by the Secretary.

(2) Every person required under this subsection to maintain records, and every person in charge or custody thereof, shall, upon request of an officer or employee designated by the Secretary, permit such officer or employee at all reasonable times to have access to and copy and verify such records.

[2]**Applicable sections of Parts F and G of Title III Public Health Service Act [42 U.S.C. 262-264] are quoted below:**

Part F - Licensing – Biological Products and Clinical Laboratories and******

Sec. 351(c) "Any officer, agent, or employee of the Department of Health & Human Services, authorized by the Secretary for the purpose, may during all reasonable hours enter and inspect any establishment for the propagation or manufacture and preparation of any virus, serum, toxin, antitoxin, vaccine, blood, blood component or derivative, allergenic product or other product aforesaid for sale, barter, or exchange in the District of Columbia, or to be sent, carried, or brought from any State or possession into any other State or possession or into any foreign country, or from any foreign country into any State or possession."

Part F – ******Control of Radiation.

Sec. 360 A(a) "If the Secretary finds for good cause that the methods, tests, or programs related to electronic product radiation safety in a particular factory, warehouse, or establishment in which electronic products are manufactured or held, may not be adequate or reliable, officers or employees duly designated by the Secretary, upon presenting appropriate credentials and a written notice to the owner, operator, or agent in charge, are thereafter authorized (1) to enter, at reasonable times any area in such factory, warehouse, or establishment in which the manufacturer's tests *(or testing programs)* required by section 358 (h) are carried out, and (2) to inspect, at reasonable times and within reasonable limits and in a reasonable manner, the facilities and procedures within such area which are related to electronic product radiation safety. Each such inspection shall be commenced and completed with reasonable promptness. In addition to other grounds upon which good cause may be found for purposes of this subsection, good cause will be considered to exist in any case where the manufacturer has introduced into commerce any electronic product which does not comply with an applicable standard prescribed under this subpart and with respect to which no exemption from the notification requirements has been granted by the Secretary under section 359(a)(2) or 359(e)."

(b) "Every manufacturer of electronic products shall establish and maintain such records *(including testing records)*, make such reports, and provide such information, as the Secretary may reasonably require to enable him to determine whether such manufacturer has acted or is acting in compliance with this subpart and standards prescribed pursuant to this subpart and shall, upon request of an officer or employee duly designated by the Secretary, permit such officer or employee to inspect appropriate books, papers, records, and documents relevant to determining whether such manufacturer has acted or is acting in compliance with standards prescribed pursuant to section 359(a)."

******

(f) "The Secretary may by regulation (1) require dealers and distributors of electronic products, to which there are applicable standards prescribed under this subpart and the retail prices of which is not less than $50, to furnish manufacturers of such products such information as may be necessary to identify and locate, for purposes of section 359, the first purchasers of such products for purposes other than resale, and (2) require manufacturers to preserve such information.

FORM FDA 482 (5/85) PREVIOUS EDITION IS OBSOLETE NOTICE OF INSPECTION

(Continued On Reverse)

FORM 3.3 - FDA Food Plant Inspection From - Rear

Any regulation establishing a requirement pursuant to clause (1) of the preceding sentence shall (A) authorize such dealers and distributors to elect, in lieu of immediately furnishing such information to the manufacturer to hold and preserve such information until advised by the manufacturer or Secretary that such information is needed by the manufacturer for purposes of section 359, and (B) provide that the dealer or distributor shall, upon making such election, give prompt notice of such election ***(together with information identifying the notifier and the product)*** to the manufacturer and shall, when advised by the manufacturer or Secretary, of the need therefor for the purposes of Section 359, immediately furnish the manufacturer with the required information. If a dealer or distributor discontinues the dealing in or distribution of electronic products, he shall turn the information over to the manufacturer. Any manufacturer receiving information pursuant to this subsection concerning first purchasers of products for purposes other than resale shall treat it as confidential and may use it only if necessary for the purpose of notifying persons pursuant to section 359(a)."

Sec. 360 B.(a) It shall be unlawful–

(1) •••

(2) •••

(3) "for any person to fail or to refuse to establish or maintain records required by this subpart or to permit access by the Secretary or any of his duly authorized representatives to, or the copying of, such records, or to permit entry or inspection, as required or pursuant to section 360A."

•••

Part G – Quarantine and Inspection

Sec. 361(a) "The Surgeon General, with the approval of the Secretary is authorized to make and enforce such regulations as in his judgement are necessary to prevent the introduction, transmission, or spread of communicable diseases from foreign countries into the States or possessions, or from one State or possession into any other State or possession. For purposes of carrying out and enforcing such regulations, the Surgeon General may provide for such inspection, fumigation, disinfection, sanitation, pest extermination, destruction of animals or articles found to be so infected or contaminated as to be sources of dangerous infection to human beings, and other measures, as in his judgement may be necessary."

FORM 3.4 – FDA Sample Collection Form - Front

DEPARTMENT OF HEALTH AND HUMAN SERVICES PUBLIC HEALTH SERVICE FOOD AND DRUG ADMINISTRATION	1. DISTRICT ADDRESS & PHONE NUMBER	
2. NAME AND TITLE OF INDIVIDUAL	3. DATE	4. SAMPLE NUMBER
5. FIRM NAME	6. FIRM'S DEA NUMBER	7. FDA'S DEA NUMBER
8. NUMBER AND STREET	9. CITY AND STATE *(Include Zip Code)*	

10. SAMPLES COLLECTED *(Describe fully. List lot, serial, model numbers and other positive identification)*

The following samples were collected by the Food and Drug Administration and receipt is hereby acknowledged pursuant to Section 704(c) of the Federal Food, Drug, and Cosmetic Act [21 U.S.C. 374 (c)] and / or Part F, Sub Part 3, Section 356(b) of The Public Health Service Act [42 U.S.C. 263d] and/or 21 Code of Federal Regulations (CFR) 1307.02. Excerpts of these are quoted on the reverse of this form.
(NOTE: *If you bill FDA for the cost of the Sample(s) listed below, please attach a copy of this form to your bill.)*

11. SAMPLES WERE ☐ PROVIDED AT NO CHARGE ☐ PURCHASED ☐ BORROWED *(To be returned)*	12. AMOUNT RECEIVED FOR SAMPLE ☐ CASH ☐ BILLED ☐ VOUCHER ☐ CREDIT CARD	13. SIGNATURE *(Person receiving payment for sample or person providing sample to FDA at no charge.)*
14. COLLECTOR'S NAME *(Print or Type)*	15. COLLECTOR'S TITLE *(Print or Type)*	16. COLLECTOR'S SIGNATURE

FORM FDA 484 (10/87) PREVIOUS EDITION MAY BE USED **RECEIPT FOR SAMPLES** PAGE OF PAGES

FORM 3.4 – FDA Sample Collection From - Rear

Section 704(c) of the Federal Food, Drug, and Cosmetic Act [21 U.S.C. 374(c)] is quoted below:

"If the officer or employee making any such inspection of a factory, warehouse, or other establishment has obtained any sample in the course of the inspection, upon completion of the inspection and prior to leaving the premises he shall give to the owner, operator, or agent in charge a receipt describing the samples obtained."

Part F, Sub Part 3, Section 356(b) of The Public Health Service Act [42 U.S.C. 263d] is quoted in part below:

"Section 356(b) In carrying out the purposes of subsection (a), the Secretary is authorized to -

(1) ****
(2) ****
(3) ****
(4) procure (by negotiation or otherwise) electronic products for research and testing purposes, and sell or otherwise dispose of such products"

21 Code of Federal Regulations 1307.02 is quoted below:

"1307.02 Application of State law and other Federal law.
Nothing in Parts 1301-1308, 1311, 1312, or 1316 of this chapter shall be construed as authorizing or permitting any person to do any act which such person is not authorized or permitted to do under other Federal laws or obligations under international treaties, conventions or protocols, or under the law of the State in which he desires to do such act nor shall compliance with such Parts be construed as compliance with other Federal or State laws unless expressly provided in such other laws."

An agreement between the Food and Drug Administration (FDA) and the Drug Enforcement Administration (DEA) provides that in the event any samples of controlled drugs are collected by FDA representatives in the enforcement of the Federal Food, Drug, and Cosmetic Act, the FDA representative shall issue a receipt for such samples on FDA form FDA 484, RECEIPT FOR SAMPLES, in lieu of DEA form 400, to the owner, operator, or agent in charge of the premises

Report of analysis will be furnished only where samples meet the requirements of Section 704(d) of the Federal Food, Drug, and Cosmetic Act [21 U.S.C. 374(d)] which is quoted below:

"Whenever in the course of any such inspection of a factory or other establishment where food is manufactured, processed, or packed, the officer or employee making the inspection obtains a sample of such food, and an analysis is made of such sample for the purpose of ascertaining whether such food consists in whole or in part of any filthy, putrid, or decomposed substance, or is otherwise unfit for food, a copy of the results of such analysis shall be furnished promptly to the owner, operator, or agent in charge."

FORM 3.5 - FDA Inspection Report Form - Front

DEPARTMENT OF HEALTH AND HUMAN SERVICES PUBLIC HEALTH SERVICE FOOD AND DRUG ADMINISTRATION	DISTRICT ADDRESS AND PHONE NUMBER	
NAME OF INDIVIDUAL TO WHOM REPORT ISSUED TO:	PERIOD OF INSPECTION	C. F. NUMBER
TITLE OF INDIVIDUAL	TYPE ESTABLISHMENT INSPECTED	
FIRM NAME	NAME OF FIRM, BRANCH OR UNIT INSPECTED	
STREET ADDRESS	STREET ADDRESS OF PREMISES INSPECTED	
CITY AND STATE *(Zip Code)*	CITY AND STATE *(Zip Code)*	
DURING AN INSPECTION OF YOUR FIRM (I) (WE) OBSERVED:		

SEE REVERSE OF THIS PAGE	EMPLOYEE(S) SIGNATURE	EMPLOYEE(S) NAME AND TITLE *(Print or Type)*	DATE ISSUED

FORM FDA 483 (5/85) PREVIOUS EDITION MAY BE USED. **INSPECTIONAL OBSERVATIONS** PAGE OF PAGES

FORM 3.5 - FDA Inspection Report Form - Rear

The observations of objectionable conditions and practices listed on the front of this form are reported:

1. Pursuant to Section 704(b) of the Federal Food, Drug and Cosmetic Act, or

2. To assist firms inspected in complying with the Acts and regulations enforced by the Food and Drug Administration.

Section 704(b) of the Federal Food, Drug, and Cosmetic Act (21 USC 374(b)) provides:

"Upon completion of any such inspection of a factory, warehouse, consulting laboratory, or other establishment, and prior to leaving the premises, the officer or employee making the inspection shall give to the owner, operator, or agent in charge a report in writing setting forth any conditions or practices observed by him which, in his judgement, indicate that any food, drug, device, or cosmetic in such establishment (1) consists in whole or in part of any filthy, putrid, or decomposed substance, or (2) has been prepared, packed, or held under insanitary conditions whereby it may have become contaminated with filth, or whereby it may have been rendered injurious to health. A copy of such report shall be sent promptly to the Secretary."

CHAPTER 4

PLANNING A PLANT SANITATION PROGRAM

The first essential for a successful plant sanitation program is for management to develop a policy statement with emphasis on producing safe and wholesome products in a clean plant using acceptable ingredients and under approved methods. The policy statement must be in writing and it must be made known to all employees. The policy statement should be of prime interest to all supervisory personnel. A typical policy statement might be as follows:

> *Our firm is committed to producing and processing high quality, safe, and nutritious products. We are successful because we care about you, our employees, and our customers whom we serve. We only use safe ingredients and we make our products uniformly. We market our products efficiently and give our customers a good value for their money. We wish to continue to grow and to that end we believe in supporting research for better raw materials and ingredients, improved processing methods, greater efficiency in processing by using better equipment and methods, and the constant development of new products for new markets. We believe in maintaining good relationships with not only our employees and our customers, but our suppliers and the public at large. We believe in giving fair and just treatment to all concerned. We anticipate continued growth and we expect all our employees to contribute to our firm's advancement.*

Management is responsible for food plant sanitation and must specifically direct an employee to be in charge of sanitation. This sanitarian must know his or her assignment exactly and management should hold him or her totally accountable for the firm's sanitation program. Management should require daily sanitation reports and back the sanitarian in his or her needs to conduct an acceptable program.

In a small plant the plant sanitarian may be one of the Quality Assurance staff members or even a production supervisor. The designated employee assigned this responsibility should have had training in microbiology, chemistry, entomology, sanitation engineering, and communications. The ideal trained person comes from one completing a BS degree in food plant sanitation, however, elected courses or even short courses can qualify a person to hold down this position. The selected individual must constantly stay up-to-date through reading, studying and attending workshops and seminars on the subject. Up-dated information is a must to accomplish the task of maintaining a plant in good sanitary condition efficiently and at reasonable costs.

The plant sanitarian should be directly responsible to top management. The plant sanitarian must be tactful and thorough in all his or her efforts. He/She must have native intelligence and the ability to communicate well with all employees on the "whys" and "hows" of food plant sanitation. He must be enthusiastic about his assignment as the sanitarian. He must always set the example on what is good sanitation by his appearance, his personality, and his attire. The following are some of the responsibilities of the plant sanitarian:

To develop an applicable plant sanitation program.
To secure the support of all personnel in the firm.
To constantly improve upon the sanitation program.
To keep informed of new developments to improve efficiency and cost reductions.

The specific duties of the plant sanitarian are as follows:

To supervise matters of personal hygiene.
The elimination of rodents and insects in and around the plant.
The maintenance of adequate cleanup of equipment and

all facilities including lunch areas and toilets.
To supervise the quality of water used in the plant, the disposal of wastes and sewage.
To supervise the proper storage of raw materials, ingredients, cleaning chemicals, food additives, and finished products.
To take corrective action where needed to prevent any product in process or in the warehouse from contamination or adulteration.
To conduct organized training programs for all plant personnel.
To make daily inspections of the firm's operation and report regularly to management.
To participate in the monthly general inspections with the management committee and follow through on taking corrective actions where needed.
To cooperate with local, state, and federal inspectors and report to management on their reports.

The plant sanitarian's tools are many and varied depending upon the type of plant and the processes involved. Generally they include the following:

Ample supply of water of the desired quality.
Nylon brushes for cleaning specific areas and equipment.
Detergents and knowledge of how to use them properly.
Chlorine and chlorination equipment for sanitizing food plant equipment.
Steam and/or high pressure equipment fitted with proper nozzles for hard to get at areas.
Vacuum cleaner with appropriate brushes for removing dirt in and around equipment.
Flashlight for inspection in out-of-the-way places.
Black light equipment for detection of rodents and molds.
Camera.
Daily record report form.
Proper attire: white helmet, white coveralls or white shirt and pants, white boots, white rubber gloves, white rubber apron, goggles.

The success of the sanitarian will be based on how well he or she is able to communicate to all employees the value and needs of operating clean equipment in a clean plant and the individual practice of proper plantkeeping.

The plant sanitation program should be organized with firm guidelines using in-plant cleaning surveys and the firm's self inspection reports. The following questions are the general areas to build a sanitation program around. Each question should be answered and if the answer is negative, the given question should be given appropriate attention to bring it into compliance for an acceptable sanitary operation and to ensure that only safe, quality products reach the consumers.

I. Plants and Grounds

1. Is the area around the buildings clear of weeds, grass and brush?
2. Are parking lots, yards, and roads adequately paved and maintained?
3. Is there any standing water in low areas which may offer insects or microbial breeding sites?
4. Is scrap, refuse, and other material stored off the ground and away from plant walls?
5. Is there any evidence of rodent burrows or insect harborage places around the perimeters of the buildings?

II. Plant Construction

6. Do windows and doors seal tightly and are they all in good repair?
7. Do open windows have fine mesh screens to keep out insects?
8. Are there any openings or cracks to provide harboring places or entry points for pests?
9. Are floors in good repair and properly sloped to eliminate water accumulations?
10. Are drains equipped with traps and are covers or grills provided and are they in good repair?
11. Are all lights in preparation and processing areas equipped with shields or proper covers?

III. Equipment

12. Is the equipment designed, or otherwise suitable, for use in a food plant?
13. Is all equipment that food comes in contact with cleaned and sanitized as often as necessary to prevent product contamination?
14. Is there any build up of food or other static materials on the equipment during operation?
15. Is the space around the equipment adequate to permit thorough cleanup?
16. Is the equipment difficult to disassemble for clean-up and inspection?
17. Are there any "dead ends" in or around the equipment where food or other debris can collect?
18. Is there any seepage of cleaning solvents or lubricants on the equipment which can contaminate the food?
19. Can the food contact surface of the equipment be sanitized?

IV. Processing And Packaging

20. Are overhead areas free of condensation from pipes and vents?
21. Are lights, equipment, pipes, and ceilings over product handling areas free of flaking paint, rust, and scale?
22. Are there any loose nuts, bolts, washers, etc. on food handling equipment?
23. Are all hand tools properly stored?
24. Is equipment stored in such a manner to prevent contami nation when used?
25. Are ingredients or materials in storage properly covered and stored between uses?
26. Is waste and trash material properly identified and removed from processing areas routinely?
27. Are cleaning materials properly identified to prevent cross contamination with processing ingredients?
28. Are incoming materials and ingredients properly sampled and analyzed to insure their quality and purity?
29. Are ingredients containers (bags, boxes, cans, bundles, glass, plastic, drums, and bulk packages) properly brushed off before usage?

V. Warehousing, Storage, Shipping And Receiving

30. Are pallets cleaned and fumigated on a regular basis?
31. Are products stored on a first in, first out basis?
32. Are products stored 18" away from walls?
33. Are spills and leaks properly cleaned up?
34. Are all products spoiled by damage, insects, rodents or other causes stored in designated "quarantine areas" to prevent their contact with safe products?
35. Are records maintained of storage area temperatures and humidity and are they within tolerances for the given products?
36. Are all incoming materials inspected and in compliance with specifications?
37. Are all outgoing materials inspected and in compliance with specifications?
38. Are receiving and shipping vehicles clean and in sanitary conditions?

VI. Employees And Employee Practices

39. Are the employees well-trained in what they are expected to do?
40. Are personal hygiene standards effectively enforced?
41. Are your employees wearing jewelry, bandages, or have any illnesses, infections or injuries which can contaminate food?
42. Are employees wearing proper hair coverings and restraints?
43. Do your employees wash their hands after each visit to the toilet or absence from their work station?
44. Are their wash stations and sanitizing areas near the employees work stations?
45. Is smoking or eating prohibited in the food preparation and processing areas?
46. Is the traffic within the plant controlled to prevent contamination of the food during processing?

VI. Toilets

47. Are toilets and urinals in good working order?
48. Are the toilets, urinals, and rest areas kept clean?
49. Is hot and cold water available for hand washing?
50. Are the toilets equipped with germicidal soap, single service towels and/or air hand dryers?

VII. Housekeeping

51. Do employees eat and smoke in designated areas?
52. Is waste and garbage removed promptly from the processing areas to appropriate covered bins?
53. Do you maintain a map showing locations of all traps and bait stations?
54. Do you check the traps and bait stations on a regular routine basis and record the findings, if any?
55. Are the baits and other poisonous chemicals for rodent, insect, bacteria, and mold control approved chemicals?
56. Are the fumigants and inhibitors approved chemicals?
57. Are chemicals (pesticides, herbicides, cleaning agents, lubricants, boiler compounds, etc.) stored in designated areas?
58. Are all hazardous materials kept in their original containers and in designated areas with appropriate MSDA sheets available to authorized personnel?
59. Are food additives stored in designated areas?

After completing the self inspection evaluation, the sanitarian should formulate a sanitation program and have it approved by management. Management has the ultimate responsibility for the sanitary conditions within and about the plant. Management must authorize the appropriate funds, personnel, and procedures for maintaining a clean plant to produce safe and wholesome foods. Good sanitation programs do not cost, they make money for a food firm.

REFERENCES

Anon. Do Your Own Establishment Inspection. A Guide to Self Inspection for the Small Food Processor and Warehouse. HHS Publication No. (FDA) 82-2163. U.S. Department of Health and Human Services, Public Health Services, FDA, Bureau of Foods. Washington D.C.

CHAPTER 5

CONSTRUCTION FACTORS

This chapter is in response to the requirements of the CURRENT CGMP's Part 110.20 which covers Grounds, Plant Construction and Design.

The grounds around a food plant should be neat, clean, and well manicured. The lawn should be well trimmed and no weeds should be permitted, thus cutting down on areas where rodents and pests may harbor. The roads leading to and from the food processing facility should be paved to cut down on dust and dirt. The parking area and/or lot should be paved to provide a satisfactory parking area for employees and visitors. The plantings in the landscaping should be maintained in a neat and clean manner with well trimmed shrubbery. All areas around the plant and facilities should be well drained to prevent any standing water whereby flies and microorganisms could breed and develop. Of particular importance are the loading and unloading dock areas. These should be kept clean and free of any debris. All efforts should be directed to protect against any possible contamination of food to be processed, in process, or finished products. Most importantly, the entrance to the plant should be attractive and inviting and give the appearance of a clean and orderly business. Ideally, the Welcome mat should always be out and pride of the community, state, and nation displayed by flying the flag(s).

Design and Layout

In the design and layout of a food plant and the construction of same, a number of considerations must be given priorities before initiating any construction:

(1) Will the plant be a Single floor layout or Multi Story Building(s)?,

(2) Will land space be made available for growth?,

(3) What consideration is given to the following:
A. - Waste Disposal,
B. - Parking for workers and visitors
C. - Access for incoming and outgoing materials,
D. - Zoning and building ordinances,
E. - Residential areas surrounding site,
F. - Safety relative environmental factors?, and
(4) Economic factors including raw materials, labor supply, energy requirements, transportation, tax structures, and community attitudes (police, fire, and government).

If space is available, a single floor facility is preferred because of cost of construction, savings in maintenance, use of straight line flow through the facility, and easier to supervise the entire operation. There are sanitation problems with single floor structures because of the additional equipment required to move the product through the facility and the excessive heights for handling the services to the equipment. Much of the conveying equipment and service areas become very difficult to keep clean.

Multi story buildings on the other hand permit gravity flow of products down through the facility and require less land area for construction. Some believe that these facilities are harder to keep clean and in a sanitary condition.

Regardless of the layout of the facility, the outer walls of the structure should be rodent proof and there should be no areas around the facility for harborage of rodents and other pests. Further the outer wall structure must be designed for ease in screening out insects.

Foundations should go to depth of three feet to prevent rats from gaining entrance and the foundation should be poured concrete or concrete blocks layed on a poured concrete foundation. If there is a cellar floor, it should be tied directly into the solid concrete foundation. However, if there is no concrete cellar floor, the foundation should include a curtain wall made of concrete and extend 2 to 3 feet below the surface surrounding the entire building with a flange at the bottom from eight to twelve inches wide extending outwards. The addition of an eight-inch flange at the bottom of a two curtain wall gives a more effective barrier than a three foot wall without a flange. The above dimensions refer to the depth of wall extending be-

beneath the ground. Temporary buildings or out-buildings can be protected by a curtain wall of galvanized metal, but it must be of identical dimensions.

Buildings with a basement concrete floor and wall present a solid concrete box. However, if the building has a poured concrete slab foundation, a flange curtain wall may be necessary to prevent rodents from getting underneath the concrete slab by digging out burrows.

Above the foundations, rodent proof materials must be used. Whenever possible, double wall construction is to be avoided. If this is not possible, then the bricks or concrete blocks must be well pointed up as to seal the double wall space to prevent rodent entry.

All **floor beams** should be installed by tying into the sills in such a way that there are no spaces left which might provide rodent or insect harborage. Where beams fit into the sill they must not provide any hollow space.

Loading docks or platforms should be built no less than three feet or truck bed height above the ground. Lower distances may permit rodents to jump onto the dock and scurry though loading

FIGURE 5.1 - Clean, well manicured outside appearence of the food plant is the first essential in food plant sanitation

doors. No stairways should lead directly from the dock to ground level. Ladder rungs should be provided in the walls for personnel use. No closed space should be left under the dock unless it is solid concrete or protected by a curtain wall such as described above for the main building foundations. There should be a 12 inch overhang on the dock to prevent rodents from access to the dock itself.

Doors, windows, skylights, transoms, ventilators and similar openings must be tight fitting and free of holes. Outside openings must be protected against the entrance of insects by using screens, fans or other suitable devices. The use of self closing doors hung so that less than 1/4 inch clearance remains when the doors are closed is mandatory for keeping mice and rodents out. Window sills and door casings should be flush with the walls on the inside of the facility and metal flashing at least 8 inches high on the outside of all outside doors is necessary for rodent proofing. If window sills are not flush with the wall, they should be at a 45 degree angle so that dust will not collect and/or no extraneous articles can be placed on them.

All windows which will be opening must be adequately screened, preferably with a No. 16 mesh screen. Screens must be tight fitting and, if possible, of metal construction designed so that new screen cloth can be readily installed as needed.

All doorways should be fitted with self closing screen doors with the exception of loading doors. The loading doors should be designed so that fly-screening fans can be installed directly above them on the outside. There must be at least forty inches of space left free above the door for the installation of the outside fly fans.

Trash and garbage dump facilities must be designed to provide for ready removal. While the trash or garbage is on the premises, it must be kept in rodent-proof containers.

Suitable drainage should be provided around the facility to allow quick run-off of any and all surface water.

The **interior structure** must be designed to facilitate cleaning. Structural features where dust accumulates should be eliminated because insects may breed there. Interior walls must be designed so that cracks and crevices do not develop. Walls must be designed so that rodents and insect harborages are not present. The walls should be smooth and easy to clean, preferably glazed tile or, if block walls, epoxy coated.

The floors should be constructed of concrete, tile, or of approved compositional material and they must extend to, and be mortise into, the structural walls on all sides. The floors and wall juncture must be curved with large enough radii to permit good sanitation. The floors must slope 1/4 inch to drains and they must be no more than 20 feet apart. Open drains should be made of one-half round 12" or larger tile and they should be covered with easily removal fitted grates for regular cleaning. All floor drains must be equipped with traps having a minimum water seal of three inches.

If pipe lines pass through the floors, ceilings, or walls, approved rodent-proof collars should cover all these openings.

Conduits, switch boxes, cabinets and other permanent wall fixtures should be either sealed in with mastic or set-out three or more inches from the wall to minimize insect harborage.

Lighting should be adequate for the job entailed. Generally, minimum lighting is no less than 20 foot candles measured at 30 inches off the floor. Inspection, office areas, and laboratories require lighting that may be as much as 150 foot candles. Proper illumination, however, is based on the size of the object to be seen, the degree of contrast between objects, the time allowed for observing the objects, and the brightness and glare of the objects. All light fixtures must be protected with plastic shields to reduce any extraneous material from broken glass in all the food items. Good illumination in a food plant is generally correlated to good plant keeping practices. Lack of good illumination is often an invitation to poor plant keeping as evidenced by the accumulation of dust and miscellaneous rubbish in the corners and out-of-the-way places. Further, studies have shown that fewer accidents occur in a well-lighted plant. In any case, illumination should be sufficient to avoid eye strain which results from marked contrasts in light intensity. All light fixtures should be regularly and thoroughly cleaned and maintained.

Of particular importance in most food plants is the ceiling in terms of design, construction, upkeep, and cleanliness. Some firms have resorted to the use of drop in "glasbord" ceiling panels using moisture and corrosion resistant Kemlite "Sanigrid" fiberglass ceiling grid systems. These grid systems are USDA accepted, they are rust and corrosion resistant, unaffected by moisture and humidity , lightweight, and easy to install. The "Kemlite Glasbord" fiberglass panels are also easy to clean, impact and stain resistant, USDA accepted, and available in various colors, including white. Further, they are Class I fire-rated.

Proper ventilation in a food plant is basic to good food plant sanitation. Control of condensation will help in the elimination of any mold growth. The use of well engineered exhaust hoods and ducts will aid in control of moisture and off-odors in the food plant. Further, positive filtered air flow in a food plant will help to eliminate dust, dirt, and most air born contaminants. All filters used in the ventilation system must be properly maintained and kept in a sanitary condition at all times.

The **plumbing** for water, waste disposal, and the steam lines must be designed to prevent any possible cross contamination. Potable water must be used for all cleaning operations, food manufacturing operations, and the personal needs of all plant personnel.

REFERENCES

Anon. Kemlite Sanigrid System info. Kemlite Company, Joliet, IL.

Gould, Wilbur A. 1977. Good Manufacturing Practices for Snack Food Manufacturers. Potato Chip/Snack Food Association, Euclid, Ohio

Imholte, Thomas J. 1984. Engineering for Food Safety and Sanitation. Technical Institute of Food Safety, Crystal, Minnesota.

Jowitt, R. 1980. Hygenic Design and Operation of Food Plant. AVI Publishing Co., Inc., Westport, Connecticut.

CHAPTER 6

EQUIPMENT DESIGN FOR FOOD PLANT SANITATION

CGMP Part 110.40 specifically states that "all plant equipment and utensils shall be so designed and of such material and workmanship as to be adequately cleanable, and shall be properly maintained." This statement is interpreted to mean the following:

1. All Food contact surfaces of equipment and utensils shall be constructed of stainless steel or other materials which are smooth, impervious, nontoxic, noncorrosive, nonabsorbent and durable under normal use conditions.
2. Food contact surfaces shall be easily cleanable, and shall be free of breaks, open seams, cracks or similar defects.
3. Food contact surfaces shall not impart any odor, color taste or adulterating substance to food.
4. Food contact surfaces shall be readily accessible for manual cleaning other than food contact surfaces designed for CIP cleaning.
5. All joints and fitting shall be of sanitary design and construction.

The equipment in a food plant is very specific for the type of foods being processed. However, there are basic factors that must be considered essential in the design and in the installation of equipment to maintain a clean plant for the production and processing of food that is clean and safe.

In the design of a food factory, the equipment should not occupy more than 20% of the floor area. Dry storage in a food plant, generally should not occupy more than 25% of the floor area, unless it is a warehouse.

Whenever possible, straight line flow of product through the plant should be utilized for ease of cleaning and supervision. Combinations or double use of specific equipment may necessitate a "T", "V", "Y", "M", or "U" arrangement of equipment.

Stainless steel, where practical, should be used for the construction of all equipment used in a food plant. Stainless steel designated at 18.8 stainless steel with a carbon content of not more than 0.08% is highly desired. It should be as smooth as 125 grit finish.

Some of the basic fundamentals of design should include the following:

1. All surfaces in contact with food should be inert to the food under the conditions of use and the food surface constituents must not migrate into the food or be adsorbed by or in the food.
2. All surfaces in contact with the food must be smooth and non-porous to the food or to bacteria, yeast and/or molds and be totally free from pits or crevices.
3. All product contact surfaces must be free of recesses, dead ends, open seams and gaps, crevices, protruding ledges, inside threads, insides shoulders, bolts and/or rivets. Permanent joints must be butt welded and the welding must be continuous, smooth, and flush with adjacent surfaces.
4. All equipment should be designed so that all contact surfaces can be readily and thoroughly cleaned and sanitized.
5. All food equipment should be available for inspection and cleaning, either by access doors, removable covers, or complete disassembly.
6. All food equipment should be designed to protect the food from external contamination including bolts, nuts, washers, gaskets, and/or other external contaminants.
7. All food equipment should be designed so as to eliminate dead ends and dead spaces or areas to harbor soil, bacteria, molds, yeasts, and other pests.
8. All food equipment should be installed with three feet clear working area around the equipment.
9. All food equipment should be installed with a minimum of six inches off the floor.

10. All moving parts should have sealed or self-lubricating bearings.
11. Hoods, if necessary, must be installed for ease in cleaning and sanitizing when appropriate.
12. Kettles and cookers should be provided with covers so designed to prevent any dripping back into the interior of the vessel and they must be self draining.

FIGURE 6.1 - Examples Of Bad Tank Design

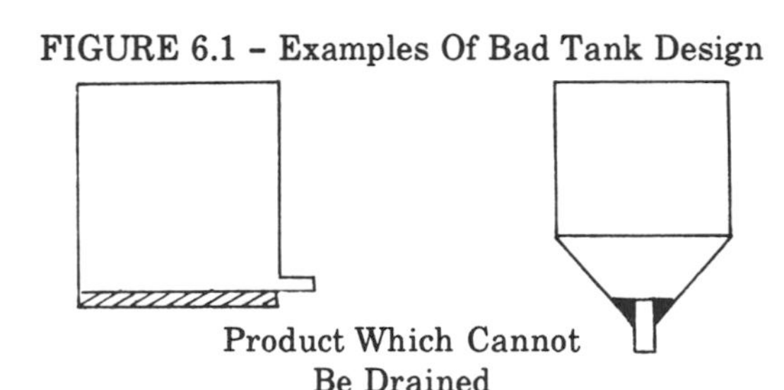

FIGURE 6.2 - Examples of Good Tank Design

13. Wood and other impervious materials shall not be used in a food plant.
14. Stainless steel should be used for the manufacture of all food plant equipment, piping, and all food contact surfaces and, where ever possible, for all supports.
15. All the food equipment should be designed with easy access for the services, preferably from above.
16. All the supports for food equipment should be of tubular steel, preferably stainless steel, with no floor flanges.
17. Whenever possible and practical, the food plant equipment should be designed for cleaning in place (CIP)

in preference to cleaning out of place (COP). In some cases the equipment may be effectively cleaned by using low pressure spray devices for soil removal.

18. All food equipment should be designed for ease in repairs due to mechanical failures.

19. All motors in a food plant should be fully enclosed, explosion and slash proof, and sealed to prevent any entrance of moisture, dust or pests.

20. No motors or drive mechanism should be mounted over the food surface areas.

21. All conveyor guides, splash guards, safety guards, etc. should be easily removed or easily opened to permit cleaning.

22. All pulleys and drums used on food equipment should be sealed or totally enclosed.

23. All water and steam valves should be designed to prevent any leakage and they should have no pockets or recess areas.

24. All food valves should be easily disassembled for cleaning and inspection.

25. Can tracks or package runways should be designed with half-round stainless steel runways or made with stainless steel rods. (See Figure 6.3 & 6.4).

FIGURE 6.3 - Modified Angle Iron Runway

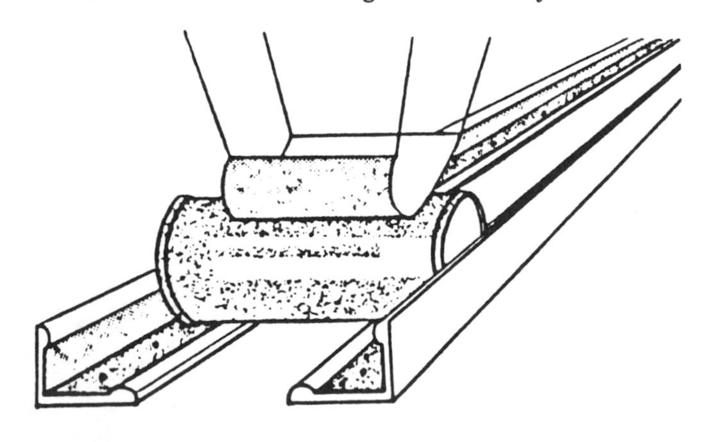

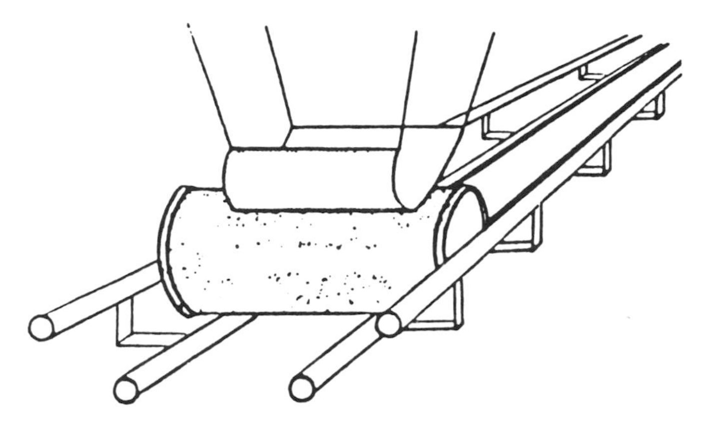

FIGURE 6.4 – Round Bar Type Open Runway

26. All piping must be aligned and supported to prevent sagging or any impediment to product flow and it must be self-draining. All corners must be provided with a minimum radii of 1 inch or more. Further, all piping must be readily removable for cleaning and inspection.
27. Piping passing through walls, floors, ceilings or other permanent structures must terminate in an accessible connection at least one foot from the wall and be rejoined to pass through. The opening in the structure must be rodent proofed.
28. All meters must be so constructed and installed that their interiors, pistons, floats, or other working parts are readily accessible for cleaning and inspection. Further all meters must be self draining and the gears or meter trains must be located outside of the product contact surfaces.
29. Angle iron, if used, as support for food plant equipment should never be used as in Figure 6.5, but it may be used as shown in Figure 6.6

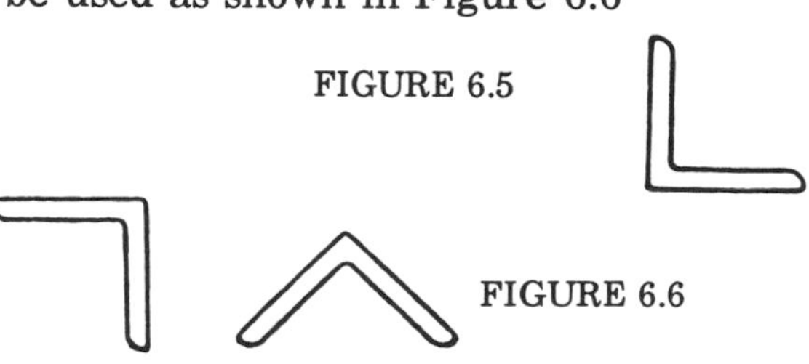

FIGURE 6.5

FIGURE 6.6

30. Channel iron, if used, in a food plant may be used as shown in Figure 6.7 or but never used as shown in Figure 6. 8.

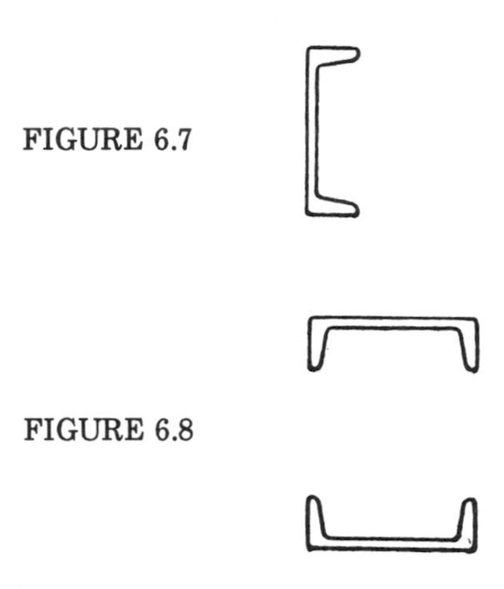

FIGURE 6.7

FIGURE 6.8

31. Iron support made into Tees are often used to support equipment and should be used as shown in Figure 6.9 but never used as shown in Figure 6.10.

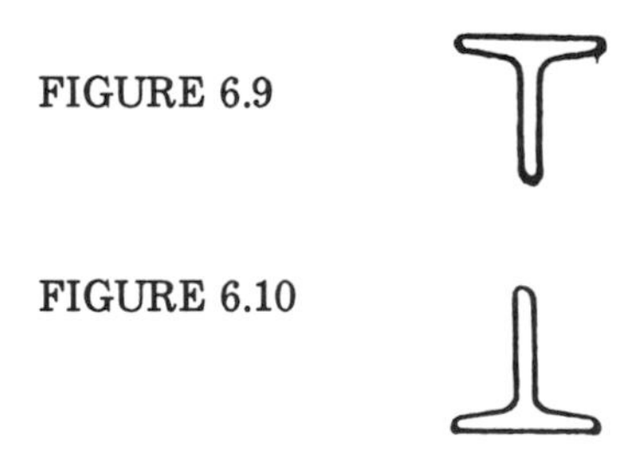

FIGURE 6.9

FIGURE 6.10

32. "I" beams should always be used in the vertical position and never flat as shown in Figure 6.11.

FIGURE 6.11

33. All instruments for measurement or control or regulating or recording processes shall be adequate in number for their designed purposes. They shall be accurate and adequately maintained.

FIGURE 6.12 - Equipment Should Be Opened Up For Inspection Purposes

REFERENCES

Banner R. 1979. Safe, Sanitary Equipment: Do you have it? Food Engineering March 111-113.

Doyle, E. G. 1965. Sanitary Design of Food Processing Equipment. J. of Milk and Food Technology. Volume 28 (10): 306-309.

Imholte, T. J. 1984. Engineering for Food Safety and Sanitation. Technical Institute of Food Safety. Minneapolis, MN.

Jowitt, R. 1980. Hygienic Design and Operation of Food Plant. AVI Publishing Co. Westport, CN.

CHAPTER 7

METAL DETECTION IN A FOOD PLANT

Upon occasions even with the best of equipment during use they may lose nuts, pieces of metal, or pieces of belt materials. Further, tramp metal may come in from the field or with ingredients. It is most important that these foreign materials be removed as they are a major source of product contamination.

The revised rule to the GCMP's now states that "effective measures shall be taken to protect against the inclusion of metal or other extraneous material in food. Compliance with this requirement may be accomplished by using sieves, traps, magnets, electronic metal detectors, or other suitable effective means (Fed. Reg., Vol. 51, No. 118, June 19, 1986).

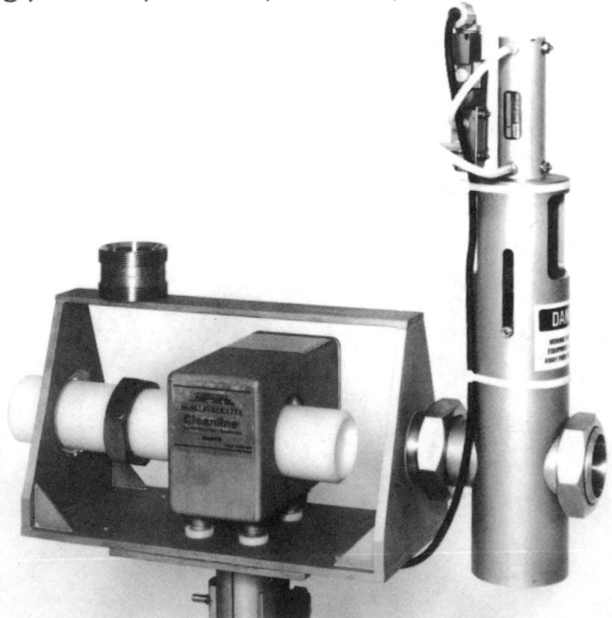

Figure 7.1 - Barkley & Dexter Model B Cleanline Metal Tracker System

The reason for detecting metal and other contaminants is to protect people from being injured and the integrity of the product. The old style metal detectors attracted only ferrous type metal (iron compounds). The detection was based on the proper positioning of the magnet and the amount of power it generated. These detectors may still have their place in the food industry. However, electronic metal detectors are quite different.

Electronic metal detectors use a high frequency electromagnetic field and metals passing through this field disturb it, that is, the metal creates a circuit imbalance and this triggers a reject system, sets off an alarm, or other appropriate signal. The advantages of this system include its ability to:

(a) Detect both ferrous and non-ferrous metals;
(b) Distinguish between product and actual metal debris;
(c) Detect metal in varying size ranges, depending on the aperture or pass-through in the systems search head; and
(d) Integrate with a product reject component ideally suited to the processing or production methods in use (CINTEX).

Installation and proper use of electronic metal detectors is one way to assure that a product is free of metal. Metal in foods not

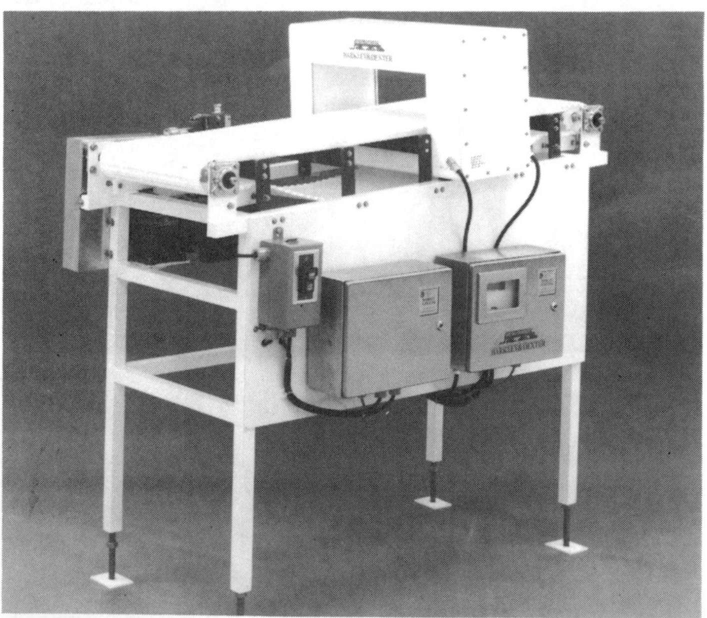

FIGURE 7.2 - Barkley & Dexter Metal Tracker Pass Through System

only may ruin a firm's reputation with resulting injury to the customer, but pieces of metal can seriously damage processing machinery, resulting in expense and production loss during the down time.

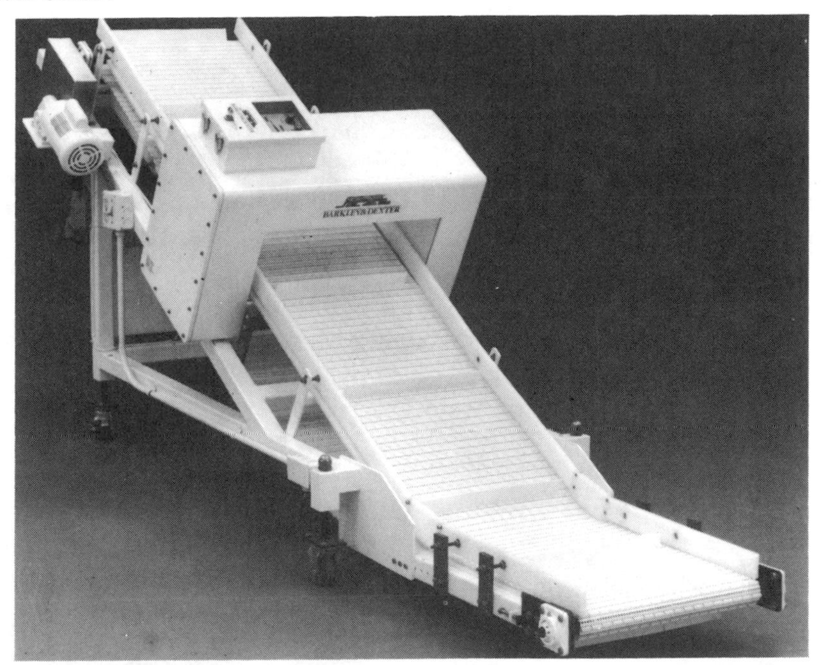

FIGURE 7.3 - Barkley & Dexter Metal Tracker Incline System

Lock International published the following list of selection of metallic contamination sources experienced by workers in this firm:

1. Raw Materials-Bulk Deliveries
   Metal clips on meat carcasses
   Staples from wooden (fiber) cases
   Wire from cutting straps
   Inefficient metal can openers
   Lead shot
2. Personnel
   Hair clips, pens, jewelry, coins
   Staples, paper clips, buttons
   Copper wire following electrical repairs
   Welding slags following repairs

3. Production and Packaging
   Broken wire sieves
   Broken mincer blades
   Swarf
   Metal slivers from grinders, cutters
   Nuts and bolts
   General wear and tear from machines, belts, shutes, etc.
   Packing materials from reclaimed product.

Regardless of the sources of contamination, it is most important that records be maintained of all contamination as a useful preventive measure. The occurrence of metal in food processing is often random and most unpredictable. Nevertheless, it is essential that food manufacturing firms make every effort to avoid any contamination through better working conditions and the use of electronic metal detectors.

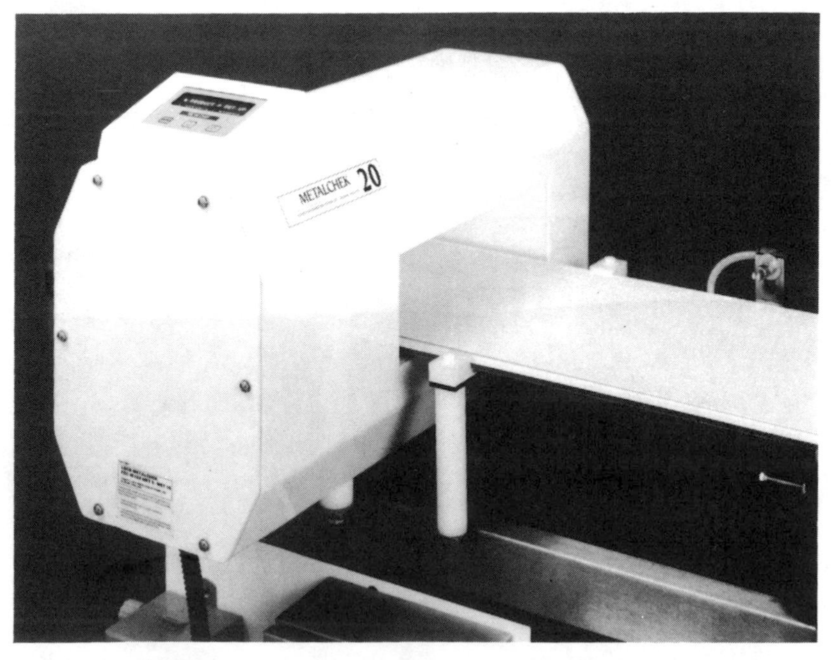

FIGURE 7.4 – Lock International Metalchek 20

It is most essential that food manufacturers use electronic metal detectors along with the following precautions (Lock International):

1. All incoming raw materials have specifications stating that the product be free of foreign bodies and that all powdered and granular products be sieved; all carcasses not be labeled with metal tags; all cans for food use be thoroughly cleaned before filling; and all pallets be in good condition and free from loose nails and broken wood, etc.
2. All bulk deliveries be received through pipelines with permanent in-line magnets.
3. De-bagging and de-boxing should be carried out in a separate area where further inspection can be conducted before transfer to covered bins. Containers should be provided for waste packaging materials.2
4. Canned products should be opened using a punch type can opener.
5. Prior to mixing, all dry ingredients should be sieved and passed through or over a permanent magnet or preferably through a free fall metal detector.
6. Liquid products should be passed through an in-line filter if practicable.
7. Watches, hair clips, decorative jewelry and similar items should not be permitted in the production areas.
8. Protective clothing should have no outside pockets.
9. Staples, thumb tacks, and paper clips should not be used on documents in production areas.
10. Magnetic "band-aid" wound dressings should be used by personnel to expedite detection of lost dressings.
11. All personnel should be properly trained and empowered to take necessary action should incidents out of order be noted, such as, defective machinery.
12. Personnel under employment termination notice should not be allowed to work in production areas since they may be less committed to taking action about incidents out of order as they may harbor resentment.

FIGURE 7.5 - Conney Plastic Strip With Metal Detectable Feature

13. All maintenance personnel and all outside contractors should be thoroughly instructed to take all precautions during production and processing operations to prevent any contamination and minimize the risk of good manufacturing practices.

14. Holding of food or ingredients should be held in covered containers until closed or capped.

15. Unfilled containers should always be inverted until filled or covered.

16. Machinery should be inspected regularly for loose or missing bolts, springs, or moving parts.

17. Equipment maintenance should not be carried out during plant operations, particularly welding and drilling.

18. When installing new lines, they should be properly screened to be certain of no contamination from construction.
19. Magnetic brushes, mats, and vacuum cleaners should be used where practical for cleanup.
20. Step bridges over open production lines should have enclosed sides and floors and be checked regularly for any possible contaminants.

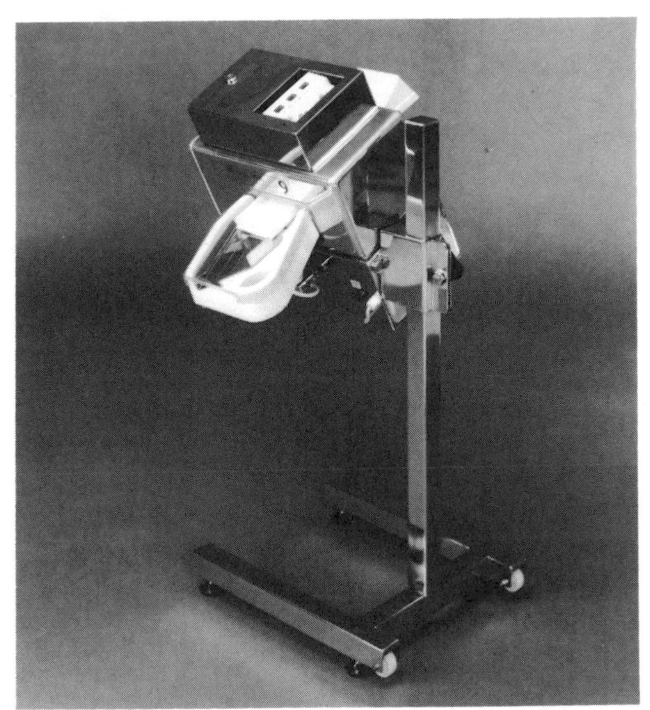

FIGURE 7.6 - Lock International Metalchek 9

The type of electronic metal detection system varies with the products being processed, type of lines and equipment being used, speeds of production, and amount of sensitivity. Food firms should deal directly with metal detection firms as to location of the detection equipment to obtain maximum benefit from the metal detection. Metal detection is one more quality assurance practice in the manufacture of safe and wholesome products.

REFERENCES

Anon. The basics of buying a metal detection system. Cintex of America, Inc., Kenosha, WI.

Anon. The Lock International Metal Detection Handbook. Yamoto Lock Inspection, Inc., Fitchburg, MA.

CHAPTER 8

WAREHOUSING, STORAGE, SHIPPING, AND RECEIVING

CGMP Part 110.93 states, "Storage and transportation of finished food shall be under conditions that will protect food against physical, chemical, and microbial contamination as well as against deterioration of the food and the container."

Sanitation applies to the warehousing, storage, shipping and receiving areas of the food business as much, if not more, than any other area in the food plant. Basic sanitary rules of construction, preparation and maintenance as previously described are all appropriate and must be followed.

All incoming materials must be sampled and evaluated to make certain they comply with the firm's specifications. Without a sound program designed to afford the plant a high assurance level in accepting merchandise, a sophisticated on-going sanitation program can be rendered unworkable. With every new shipment the assurance level of the warehouse is threatened, unless a sound and practical system of acceptance and rejection is established. By adopting a careful system of acceptance of all merchandise into the plant, all vendors become aware that the firm will not accept merchandise that does not meet specifications. Accepting a contaminated or infested lot of raw materials or other ingredients is the fastest and surest way to spread infestation throughout the warehouse. In addition all returnable cartons must be thoroughly inspected upon receipt of same.

The receiving department has the responsibility to inspect all loads of merchandise including raw materials coming into the plant or warehouse. They must be held responsible for thorough inspection and making the decision of acceptance or rejection. If any merchandise is contaminated or suspected of being contaminated, the load or lot shall be segregated and either adequately reconditioned or rejected.

The receiving inspector must be trained to look for external evidences of:

1. Live insects crawling or insect trails on the product, in the product containers, or in the containers themselves.
2. Old products that may be potential generators of infestation.
3. Fruit flies or other flying insects.
4. Toxic materials or other chemical substances in the receiving containers.
5. Presence of rodent urine or pellets, gnawing, nests, or live animals or carcasses.
6. Bird excreta.
7. Undesirable odors.
8. Stains or dust on product containers.
9. Shipping containers, pallets, or rolling equipment interiors that may be excessively dirty or food spattered.
10. Other possible contaminants.

The inspector should complete a record form on every lot of merchandise received in the plant. The record should show the number of samples tested, the size of the load, and the specific evaluation data indicating any of the conditions enumerated above. Most firms go one step further and evaluate product quality of all incoming raw materials as to acceptance levels of quality. All vendors should be rated on their promptness of delivery as well as the quality of merchandise.

Products in the warehouse must be stored on clean pallets and, at least, 4 inches off the floor and 18 inches from outside walls. All warehouse stock should be maintained in a clean and orderly manner at all times. No odorous materials, hazardous chemicals, insecticides, or animal feeds shall be stored in close proximity to food items.

All warehouse stocks should be on a rotation system. If incoming products are not coded, the warehouseman should set up a code system including date of receipt. The warehouseman should practice First In First Out (FIFO).

Set aside areas should be designated for damaged or rework products. Any torn containers should be taped up or otherwise closed to prevent entrance of any possible contamination and to prevent further spillage.

The floors in a warehouse should be kept clean and neat at all times. Any oil leaks from fork lift trucks or spills should be carefully cleaned up.

Doors and windows should be kept closed except for entry and egress to prevent birds and rodents from entry.

Bait boxes, if needed, must be carefully monitored on a regular basis and damaged bait boxes must be removed and all bait carefully cleaned up.

If the warehouse must be fumigated, licensed control operators must be used and they must carefully follow directions for any fumigants used in food areas.

Prior to loading any vehicle for transport of food products, a thorough inspection of the interior should be made to be certain that it is clean, free from moisture, and free from materials which could cause product contamination or damage to the products. If the vehicle is not cleaned and sanitary, it should be cleaned prior to loading.

In addition to warehouse or storage areas, constant inspection should be made of the loading/unloading docks, siding, truck bays and driveways to be certain that there is no accumulation of debris or spillage or no harborage of insects, birds or rodents. These areas should never be neglected as they are the ideal entrance areas. All holes, crevices, and cracks must be kept in repair.

The Food and Drug Administration has published rules for food warehousemen as follows:

1. Promote personal cleanliness among employees.
2. Provide proper toilet and hand-washing facilities.
3. Adopt "good housekeeping" practices.
4. Keep food handling equipment clean.
5. Reject all incoming contaminated foods.
6. Maintain proper storage temperature.
7. Store food away from walls.
8. Rotate stock and destroy spoiled foods.
9. Do not use or store poisonous chemicals near foods.

10. Maintain an effective pest control program:
    a. Assign inspection and reporting duties to a dependable employee.
    b. Keep building insect, bird and rodent proof.
    c. Keep doors closed when not in use.
    d. Follow label directions exactly when applying insecticides and rodenticide.
    e. Use highly toxic rodenticide only in locked bait boxes.
    f. Remove and prevent litter around buildings.
    g. Be alert for signs of rodents and insects.

Incoming materials into a food plant are the first major source of food plant contamination. All incoming materials, carriers, containers and pallets may be sources of contamination. The Food and Drug Administration has developed the following material, procedures and report forms as a guide for those receiving materials in a food plant. It is presented here to help you assure your firm of clean, safe and wholesome incoming materials.

Why Should You Inspect Incoming Food Materials?

a. To make money, your firm must handle only good products. Rotten, spoiled or contaminated food materials will never change into good products!

b. Very often, firms which accept contaminated and spoiled food materials are *forced to go out of business.* When this happens, the employees lose their jobs!

c. Contaminated and spoiled food can make people sick.
Since most of us cannot be there to inspect the food materials as they are delivered, *consumers depend on you* to make a good inspection and to make sure contaminated food materials do not enter your plant.
To do this right, you have to know how to Inspect and what is needed to make a good Inspection.

Before The Shipment Arrives:
Make Sure That:
1. The storage space for the shipment is clean and dry;
2. The equipment you will use to handle incoming food materials is clean and in good repair;
3. You have the following tools so you can make a good inspection:
   - a magnifying glass
   - flashlight
   - black light (ultraviolet light) source (for identifying rodent urine)
   - sample containers (plastic bags with self seals or glass jars with covers)
   - sample thief, trier and spatula
   - other equipment to aid inspection of specific products
   - Inspection Report form
   - marking pencil
4. You do not contaminate the product during sampling
5. You also follow specific instructions given by your supervisor.

If you follow these and your supervisor's instructions and use the equipment properly, you will make a good inspection and help assure that only clean, wholesale ingredients and food materials are used in the products you help manufacture.

GUIDELINES FOR INSPECTIONS

Note outside condition of carrier
The outside condition of the carrier may indicate contents were exposed to contamination while in transit.

1. Mud, dirt, water, oil stains (Photo a) or heavy insect debris on outside of carrier may have found its way to the products. For example, if the outside is wet, seepage may have occurred and contaminated the contents.

2. The shipment is more likely to be contaminated if the carrier is (1) an open-bed truck that is not properly covered; or (2) a truck or boxcar that is visibly damaged (photo b).

a.

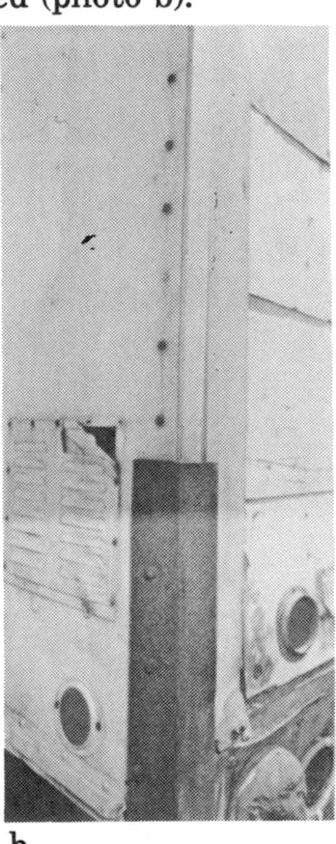

b.

Notify your supervisor if you suspect shipment was exposed to contamination while in transit. Also note in your inspection report the type of carrier bringing the shipment.

If doors of vehicle compartments have a seal, note if it is broken.

The manufacturer affixed the seal to assure that you receive the high quality products he manufactured and shipped; if the seal is broken the acceptability of the products in the shipment should be suspect.

1. A broken seal (Photo b) may indicate that some of the merchandise was stolen or that poor quality products may have been substituted after your shipment was loaded and before arrival at your plant.

2. Toxic non-food items may have been added to the load, possibly contaminating your products, and removed before delivery of your shipment.

3. Compartment doors may have been opened to air-out foul odors shortly before arrival at your receiving dock. (Odors may have accumulated from trash, filth or spillage from previous shipment or your present shipment.)

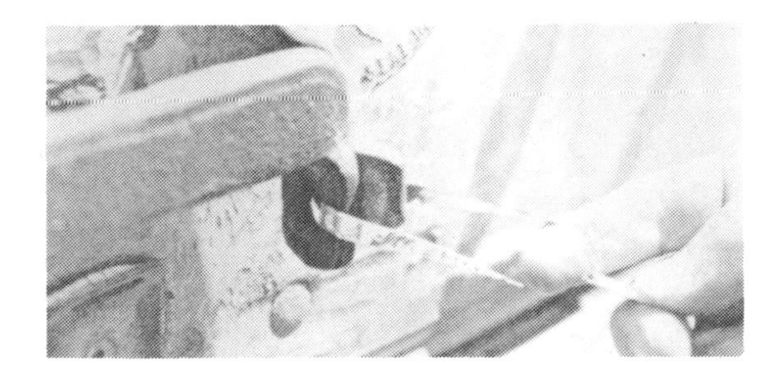

a.

b.

Do not accept shipment if seal is broken notify your supervisor before proceeding further with the inspection and receiving.

Break seal and open doors -
at the same time note odor and temperature.

Check for off odors, and high temperatures (Photo a). If you find off odors in any shipment, or the temperature is high in refrigerated loads, it may mean the delivered products are unsafe.

1. Foul odors may have been caused by the failure to remove *food particles, filth and infestation* resulting *from previous shipments* and to *clean the carrier properly* before loading your shipment.
2. The *products* may have been decomposed before being loaded, causing the off-odor.
3. The *products* may have absorbed *harmful off-odors before shipment.*
4. *Toxic solvents, petroleum products or chemicals* may have been carried with your shipment and unloaded before arrival at your receiving dock.
5. *Frozen products,* in the refrigerated load, may have been allowed to thaw (Photo b) during shipment, permitting the bacteria to grow and produce off-odors which accumulated.
6. *High temperature,* in refrigerated compartments, will allow the few bacteria normally present in the products to *increase to dangerous numbers* and to *produce harmful decomposition products and odors* (Photo c).

a.

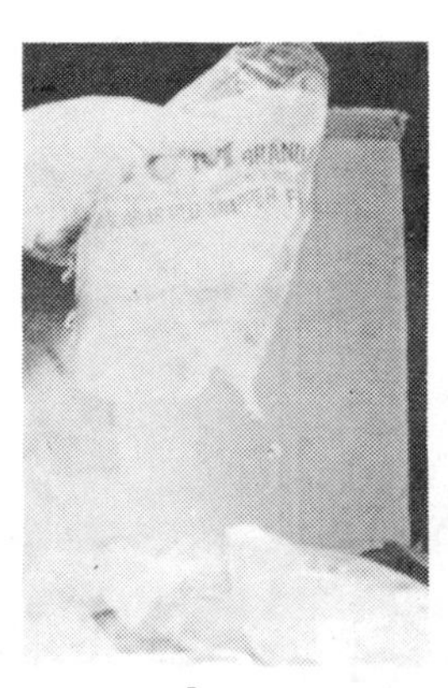

b.

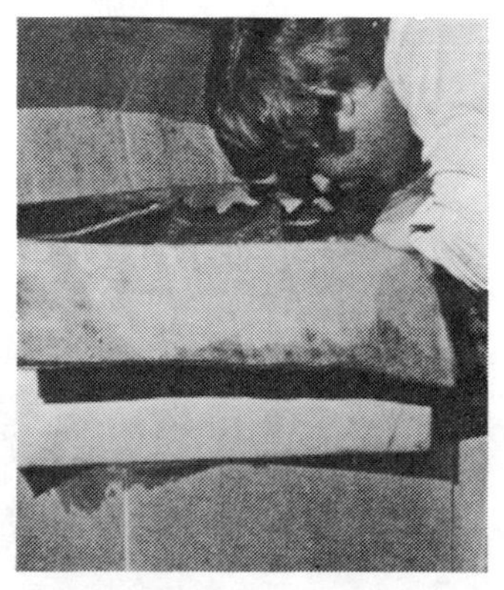

c.

Do not accept shipment if off-odor or high temperature is observed. Instead, close compartment doors immediately and tell your supervisor. Such products can be a danger to health therefore may be seized.

Note condition of stacked cartons or other containers

Packages, cartons and similar type containers protect the products they contain (Photo a). If they are broken, crushed or otherwise damaged, their contents will be exposed to possible contamination.

1. It is difficult to prevent contamination of food products in damaged packages, cartons or other containers.

2. Broken packages or containers (Photo b) may mean the product was contaminated and violative before it was loaded and shipped. The damage may have occurred while the product was in storage and contents exposed to insects, rodents or other contamination while awaiting shipment.

3. Harmful chemicals or pesticides may have entered the broken containers.

4. The shipment may have been improperly stacked or mishandled while loading or not protected while enroute to your plant.

a.

b.

c.

Do not invite trouble – set aside all damaged cartons, containers and packages (Photo c). Do not tape over or repair holes or other damages you may find in packages or cartons; report to your supervisor if you discover many broken or damaged cartons.

Look for evidence of insect, rodent and bird activity

Finding the presence of insects (Photo a), rodent excreta bird feathers or droppings or rodent urine (Photo b- using ultraviolet light) is evidence products were exposed to contamination making them unfit for food. Do not accept shipments containing insect, rodent, or other filth.

1. Insects, rodents and birds are often carriers of disease-producing bacteria and parasites. Rodent excreta (Photo c) or droppings and urine can transfer these organisms to food products.

2. Products may have been contaminated with this filth before being shipped to your plant.

3. FDA will seize products stored in your warehouse if they are exposed to or contain insect or rodent or other filth. The filth does not have to be found in exposed products to make the product subject to legal actions.

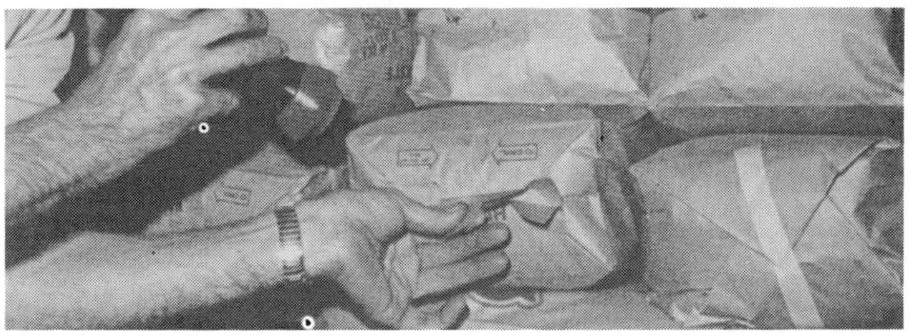

a.

b.

c.

Notify your supervisor as soon as possible when you find evidence of insect, rodent, bird or other contamination in the shipment.

Remove random samples of food containers for product examination

Random samples should be collected from the shipment and examined for contamination either on-the-spot or in the laboratory.

1. It is not possible and not practical to examine the contents of every packaged product in the shipment because the package is not saleable after opening and may become contaminated before being used.
2. Random samples that are representative of those in the entire shipment can be relied upon to show if products are acceptable or contaminated.
3. We can get a true picture of the entire lot ONLY if the samples are collected RANDOMLY (that is, every 10th, 12th, 30th or whatever, depending on the number in the shipment).

   If you are given the job to unload and inspect the shipment and no one is available for on-the-spot examination of the contents of packages, *ask your supervisor for instructions (Photo a) as to the number of cartons of packages he wants you to take randomly from the load* to set aside for later examination either on-the-spot or in the laboratory. *Follow his instructions carefully* because it is important that samples be collected randomly (Photo b).

   If you are assigned to make on-the-spot examinations of collected samples, *Be sure you follow proper instructions and know how to use all of the inspection tools listed in the front of this booklet. Ask your supervisor for more specific instructions for on-the-spot or laboratory examinations.*

a.

b.

Proper sample collection and examination will help prevent accepting contaminated shipments that should be rejected. Do your part to help your supervisor make the proper decision.

While unloading, note if non-food items are also in the shipment

Write in your report any non-food products (liquid or dry) that you find in the shipment with the food items.

1. Such products may be poisonous and can be a source of food product contamination. For example: there is no way you can be sure containers of non-food products will Not Leak or Break (Photo a) during shipment or storage and contaminate the food items, making them POISONOUS OR OTHERWISE UNFIT TO EAT which can happen without your knowing it since NO CHANGE IN THE LOOKS OF THE FOOD PRODUCTS MAY TAKE PLACE.

2. YOU DON'T HAVE TO FIND THE POISONOUS STUFF IN THE MATERIAL TO REJECT THE SHIPMENT. The Food, Drug and Cosmetic Act, which protects consumers, says in simple words that a food product is illegal if it is prepared, packed or held (or shipped) under conditions which may have caused it to become contaminated with filth or which may have caused it to become dangerous to the of consumers (such as by exposing it to poisonous substances).

a.

Be very careful about accepting foods shipped or stored with non-food products that may be poisonous.

After unloading, observe inside condition of carrier

Floors and walls in disrepair (Photo a) and residue wastes from non-food shipments can cause contamination.

1.Cracks and broken boards are good hiding places for insects which could invade the shipment while in transit.

2. Residues from non-food items previously shipped in the carrier can contaminate food products (Photo b).

3. Presence of cracks, splinters or broken boards may have prevented satisfactory cleaning and sanitizing of the carrier's interior, prior to loading your shipment, increasing the chance for contamination.

a.

c.

b.

If the inside condition of the carrier you are inspecting is bad, mention it in your inspection report. To discourage infestation, make sure all of the paper liners and wastes from your shipment are removed and the truck or railcar is swept clean (Photo c) before releasing it.

FORM 8.1 – Incoming Food Materials Inspection Report

INCOMING FOOD MATERIALS

Inspection Report

Shipment/Product________________ DateInspected________________

Carrier ________________ Name of Inspector ________________

To the Inspector: Check the space/spaces which indicate what you found during your inspection. Also note in the "Remarks" section anything else you found which is not listed on the form.

1. Is the outside of the carrier:
☐ clean ☐ muddy ☐ dirty ☐ oily ☐ other (describe)________________

2. Is the compartment door seal(s): ☐ OK ☐ broken ☐ missing

3. As you open the compartment door(s):
Does it smell clean - ☐ yes ☐ no
Do you smell off-odors - ☐ yes ☐ no
petroleum distillate - ☐ yes ☐ no
putrid or sour - ☐ yes ☐ no
other (describe) ________________

4. Are boxes, cartons or containers:
properly stacked - ☐ yes ☐ no
☐ crushed ☐ broken ☐ scattered

5. Is there evidence of activity by:
☐ insects (live or debris)
☐ rodents (pellets, urine)
☐ birds (droppings, feathers)
(For example, is there evidence of nesting in cracks, corners or inside broken partitions; bird droppings, insect debris or rodent pellets or urine stains on containers, walls or floors.

IF YES TO #5, NOTIFY YOUR SUPERVISOR IMMEDIATELY

6. Is there evidence of misuse of pesticides such as:
DDT tracking powder - ☐ yes ☐ no
1080 - ☐ yes ☐ no
insect sprays - ☐ yes ☐ no
other (describe) ________________

IF YES TO #6, NOTIFY YOUR SUPERVISOR

FORM 8.1 – Incoming Food Materials Inspection Report - Continued

7. Does the shipment include harmful non-food items? ☐ yes ☐ no
Describe ____
(Examples: chemicals, solvents, pesticides, cleaning compounds, or other items)

IF YES TO #7 NOTIFY YOUR SUPERVISOR IMMEDIATELY

8. Sampling and examination information:
Number of cartons in shipment __________
Number of cartons randomly selected/set aside for examination ________
Number of cartons opened __________
Number of packages taken from cartons that were examined _________
Number of packages found contaminated __________

9. Condition of inside of carrier:
in: ☐ excellent shape ☐ good shape ☐ poor condition
☐ damaged ☐ dirty - describe ____________________________________
☐ infested (describe) ____
☐ contains trash from previous shipment ☐ yes ☐ no
☐ has been swept clean

10. Additional comments/remarks/observations that you may wish regarding what you observed during the inspection.

____

____

____

RECOMMENDATION: ☐ Accept ☐ Reject* - this shipment

Remember - A good thorough inspection of incoming food materials is the first line of defense against producing infested or otherwise contaminated finished products. By following the guidelines of this inspection report and recording your findings, you will make a good beginning and will greatly help your supervisor to make the correct decision regarding rejection or acceptance of incoming shipments.

**Many firms provide additional consumer protection by notifying the local FDA office regarding shipments they have rejected.*

CHAPTER 9

OTHER FACILITIES AND FOOD PLANT SANITATION

Several other facilities are important to food plant sanitation. The most important of these are the toilets, lockers, lunch rooms, office areas, and the maintenance areas. Most of these areas are part of the employee's personal needs. People are the most significant part of the food firm and their needs are important for them to have pride in their place of work. Further, the attitude people display about their work place is carried over from the facilities available to them.

Adequate toilet facilities, conveniently located, must be furnished for comfort of the employees. This area should be revised as the number of employees changes to keep in line with basic requirements as set forth in the Occupational and Safety Health Act of 1970 (see back of chapter). Ample toilet facilities means sufficient stools, urinals, lavatories, soap, drying facilities and toilet tissues dispensers, cigarette urns, waste receptacles, and mirrors.

The floors, walls, ceilings, partitions, and doors of all toilet rooms shall be of a finish that can be easily cleaned.

All water closets shall have a hinged seat made of substantial material having a nonabsorbent finish. All seats shall be open front type. The toilet bowl should be set entirely free from the wall and partitioned, so that the space around the fixture may be cleaned easily.

All toilet facilities should be partitioned off and the compartment partitions should be no less than six feet high and should be one foot off the floor.

The floors should be constructed with a watertight, non-absorbent material, sloped to a central drain, and provided with a coved base extending up the wall to height of at least 5 inches.

Doors should be made self-closing and the doorway screened, so as to conceal the inside of the toilet room from the outside.

Toilet rooms should be always maintained, well lighted, and ventilated to outside air and all doors, windows, or other opening equipped with fly screening.

Floors, toilets, and urinals must be cleaned daily.

At least one covered receptacle should be provided in every toilet room used by females.

Hand washing signs must be conspicuously posted in toilet and lavatory rooms and personnel must wash their hands after the use of the facilities. Further, all employees should wash their hands before starting work, after each absence from the work area or at other times when their hands have become soiled or contaminated. At least one wash basin or equivalent should be provided for each ten persons up to 100 persons, and one for each fifteen people over 100 (24 inches of sink, with faucet, or circular basin of 20 inches may be considered equal to one individual basis). Both hot and cold water should be available from the same mixing faucet. Soap, in a separate dispenser, must be provided at each basin.

FIGURE 9.1 - Meritech CleanTech™, Model 2000S Hand Sanitation System

Individual hand towels or sections thereof of cloth or paper or warm air blowers shall be provided convenient to the washing areas. Receptacles shall be provided for disposal of the used paper towels. The warm air blowers shall provide air at not less than 90° F and shall have means to automatically prevent the discharge of air exceeding 140° F

Some firms are installing MERITECH Hand Sanitation CleanTech Automatic Systems to wash and sanitize hands. The free standing wash station is operated by placing the hands through two ports. An optical sensor activates the wash cycle, which is pre-set at 10 seconds, but can be adjusted to 6-20 seconds. A half gallon of water is required for the complete 10 second cycle, which also uses 2% chlorohexidine gluconate as a disinfectant, with a residual kill capability for up to six hours. Water and cleansing solution are sprayed onto the hands, from finger tips to wrists. This system is a sure way of making certain employees are washing and sanitizing their hands. The system is, also, effective to make certain that gloves are free of contaminates.

A rest room must be provided when ten or more women are employed. The minimum floor area is 60 feet plus 2 additional feet for each additional woman over 10 in number. One bed is required for 10 to 100 women, two beds with 100 to 250 women and 1 additional bed for each additional 250 women. Suitable bed linen must be provided and maintained in a clean condition.

Whenever showers are required, a shower shall be provided for each 10 employees of each sex, or numerical fraction thereof, who are required to shower during each shift. Body soap or other appropriate cleansing agents convenient to the showers shall be provided. Showers shall be provided with hot and cold water feeding a common discharge line. Any employees required to use the showers shall be provided with clean individual towels.

Each plant employee who is required to change clothes in order to perform his or her duties is entitled to an individual locker and the accountability for keeping the locker clean rests with this employee. The responsibility for holding the individual employee accountable for keeping the lockers clean, however, rests with the supervisor. The best way to exercise this responsibility is to set a regular schedule for locker cleanout every 3 or 4 weeks. Even though a locker is assigned to an employee, it still remains company property and must be kept clean and sanitary as a company responsibility.

Lockers should be free standing or mounted on walls one foot up from the floor. They should be sealed to prevent harborage of any pests. The top of the locker should be sloped to prevent storage of items on top and to prevent accumulation of dust etc. The lockers should of metal construction as well as stools or benches provided for changing of clothes. Plastic may be substituted where appropriate.

Lunch rooms must be equipped with washable furniture. The lunch room should be pleasant and decorated with relaxing colors with appropriate lighting. Dispensing equipment must be kept clean, at appropriate temperatures, and in a sanitary condition. Proper waste disposal receptacles shall be provided and used for the disposal of food wastes. The furniture, floors, and dispensing equipment or other food facilities shall be kept clean and in a sanitary condition at all times. Recommended areas for lunch facilities include: 13 sq. ft. per person for 25 or less employees, 12 sq. ft. for 26 to 74 employees, 11 sq. ft. for 75 to 149 employees and 10 sq. ft. for 150 or more employees.

The maintenance area should always reflect all the principles of good sanitation and good housekeeping practices as evidenced by proper storage, orderliness and cleanliness of tools, equipment, and machinery in repair. The maintenance supervisors desk should be used for clerical purposes and not as a work bench. The plant maintenance personnel should always be proud of the condition of the ship and be able to use it as an example for other departments to follow.

Like wise the boiler rooms should be kept clean and free of miscellaneous items of storage. This room should be a part of the regular inspections.

All entrances, stairways, and halls serving the general public and employees should represent the finest of efforts in maintenance and cleanliness at all times.

The plant office should represent the ultimate in cleanliness and orderliness at all times. Floors, desks, cabinets, and the like should be cleaned and polished regularly. Light fixtures, glass partitions, windows, and the like should be washed regularly and otherwise kept clean. Of course, the walls and ceilings must be kept clean and appropriately painted. At the close of each day, all papers, books, forms, etc. should be returned to the appropriate files or storage cabinets, thus leaving the desks and furniture free of all paper work to permit the areas to be properly cleaned daily.

An adequate supply of cool, potable drinking water must be provided at convenient places on each floor of the plant with one fountain for each 50 employees. Portable drinking fountains or water dispensers shall be designed, constructed, and serviced so that sanitary conditions are maintained. Any ice used in a drinking fountain or for employees shall be made of potable water and maintained in a sanitary condition. Drinking facilities shall not be placed in the toilet rooms. Proper receptacles shall be provided for disposal of single service drinking cups.

Sanitation in a food plant can be easily sold if the personal conveniences of the employee are provided and properly maintained. Good will toward employees must be created by management by setting the right example in these facilities.

OSHA Sanitation Standards

Accordingly, after consideration of all materials presented and pursuant to Sections 6(b) and 8(g) of the Williams-Steiger Occupational Safety and Health Act of 1970 (29 U.S.C. 655, 657). Secretary of Labor's Order No. 12-71 (36 FR 8754), and 29 CFR 1911.5, 29 CFR 1910.141 is hereby revised effective June 4, 1973, to read as follows:

Section 1910.141 Sanitation

(a) General

(1) Scope This section applies to permanent places of employment.

(2) Definitions applicable to this section:

(i) **Lavatory** means a basin or similar vessel used exclusively for washing of the hands, arms, faces, and head.

(ii) **Nonwater** carriage toilet facility means a toilet facility not connected to a sewer.

(iii) **Number of employees** means, unless otherwise specified, the maximum number of employees present at any one time on a regular shift.

(iv) **Personnel service room** means a room used for activities not directly connected with the production or service function performed by the establishment. Such activities include, but are not limited to, first-aid medical services, dressing, showering, toilet use, washing, and eating.

(v) **Potable water** means water which meets the quality standards prescribed in the U.S. Public Health Service Drinking Water Standards, published in 42 CFR part 72, or water which is approved for drinking purposes by the State or local authority having jurisdiction.

(vi) **Toilet facility** means a fixture maintained within a toilet room for the purpose of defecation or urination or both.

(vii) **Toilet room** means a room maintained within or on the premises of any place of employment containing toilet facilities for use by employees.

(viii) **Toxic material** means a material in concentration or amount which exceeds the applicable limit established by a standard such as sections 1910.93 and 1910.93a or in absence of an applicable standard, which is of such toxicity so as to constitute a recognized hazard that is causing or is likely to cause death or serious physical harm.

(ix) **Urinal** means a toilet facility maintained within a toilet room for the sole purpose of urination.

(x) **Water closet** means a toilet facility maintained within a toilet room for the purpose of both defecation and urination and which is flushed with water.

(xi) **Wet process** means any process or operation in a workroom which normally results in surfaces upon which employees may walk or stand becoming wet.

(3) Housekeeping

(i) All places of employment shall be kept clean to the extent that the nature of the work allows.

(ii) The floor of every workroom shall be maintained so far as practicable in a dry condition. Where wet processes are used, drainage shall be maintained and false floors, platforms, mats, or other dry standing places shall be provided, where practicable, or appropriate waterproof footgear shall be provided.

(iii) To facilitate cleaning, every floor, working place, and passageway shall be kept free from protruding nails, splinters, loose boards and unnecessary holes and openings.

(4) Waste disposal

(i) Any receptacle used for putrescrible solid or liquid waste or refuse shall be so constructed that it does not leak and may be thoroughly cleaned and maintained in a sanitary condition. Such a receptacle shall be equipped with a solid tight-fitting cover, unless it can be maintained in a sanitary condition without a cover. This requirement does not prohibit the use of receptacles which are designed to permit the maintenance of a sanitary condition without regard to the aforementioned requirements.

(ii) All sweepings, solid or liquid wastes, refuse, and garbage shall be removed in such a manner as to avoid creating a menace to health and as often as necessary or appropriate to maintain the place of employment in a sanitary condition.

(5) Vermin Control - every enclosed workplace shall be so constructed, equipped, and maintained as to prevent the entrance or harborage of rodents, insects, and other vermin. A continuing and effective extermination program shall be instituted where their presence is detected.

(b) Water supply -

(1) Potable water

(i) Potable water shall be provided in all places of employment for drinking, washing of the person, cooking, washing of foods, washing of cooking or eating utensils, washing of food preparation or processing premises, and personal service rooms.

(ii) Drinking fountain surfaces which become wet during fountain operation shall be constructed of materials impervious to water and not subject to oxidation. The nozzle of the fountain shall be at an angle and so located to prevent the return of water in the jet or bowl to the nozzle orifice. A guard shall be provided over the nozzle to prevent contact with the nozzle by the mouth or nose of persons using the drinking fountain. The drain from the bowl of the fountain shall not have a direct physical connection with a waste pipe, unless it is trapped.

(iii) Portable drinking water dispensers shall be designed,constructed, and serviced so that sanitary conditions are maintained, shall be capable of being closed, and shall be equipped with a tap.

(iv) Ice in contact with drinking water shall be made of potable water and maintained in a sanitary condition.

(v) Open containers such as barrels, pails, or tanks for drinking water from which the water must be dipped or poured, whether or not they are fitted with a cover, are prohibited.

(vi) A common drinking cup and other common utensils are prohibited.

(vii) Where single service cups (to be used but once) are supplied, both a sanitary container for the unused cups and a receptacle for disposing of the used cups shall be provided.

(2) Nonpotable water

(i) Outlets for nonpotable water, such as water for industrial or firefighting purposes, shall be posted or otherwise marked in a manner that will indicate clearly that the water is unsafe and is not to be used for drinking, washing of the person, cooking, washing of food, washing of cooking or eating utensils, washing of food preparation or processing premises, or personal service rooms, or for washing clothes.

(ii) Construction of nonpotable water systems or systems carrying any other nonpotable substance shall be such as to prevent backflow or backsyphonage into a potable water system.

(iii) Nonpotable water shall not be used for washing any portion of the person, cooking or eating utensils, or clothing. Nonpotable water may be used for cleaning work premises, other than food processing and preparation premises and personal service rooms, provided that this nonpotable water does not contain concentrations of chemicals, fecal coliform, or other substances which could create insanitary conditions or be harmful to employees.

(c) Toilet facilities -

(1) General

(i) Except as otherwise indicated in this subdivision (1), toilet facilities, in toilet rooms separate for each sex, shall be provided in all places of employment in accordance with Table 9-1 of this section. The number of facilities to be provided for each sex shall be based on the number of employees of that sex for whom the facilities are furnished. Where toilet rooms will be occupied by no more than one person at a time, can be locked from the inside, and contain at least one water closet, separate toilet rooms for each sex need not be provided. Where such single-occupancy rooms have more than one toilet facility, only one such facility in each toilet room shall be counted for the purpose of Table 9-1.

Where toilet facilities will not be used by women, urinals may be provided instead of water closets, except that the number of water closets in such cases shall not be reduced to less than the minimum specified.

(ii) The requirements of subdivision (1) of this subparagraph do not apply to mobile crews or to normally unattended work locations so long as employees working at these locations have transportation immediately available to nearby toilet facilities which meet the other requirements of this subparagraph.

TABLE 9.1 - Minimum Number Of Water Closets

Number of employees	Minimum number of water closets
1-15	1
16-35	2
36-55	3
55-80	4
81-110	5
111-150	6
Over 150	Additional fixture for each additional 40 employees

(iii) The sewage disposal method shall not endanger the health of employees.
(iv) When persons other than employees are permitted the use of toilet facilities on the premise, the number of such facilities shall be appropriately increased in accordance with Table 9-1 of this section in determining the minimum number of toilet facilities required.
(v) Toilet paper with holder shall be provided for every water closet.
(vi) Covered receptacles shall be kept in all toilet rooms used by women.
(vii) For each three required toilet facilities at least one lavatory shall be located either in the toilet room or adjacent thereto. Where only one or two toilet facilities are provided at least one lavatory so located shall be provided.

(2) Construction of toilet room
(i) Each water closet shall occupy a separate compartment with a door and walls or partitions between fixtures sufficiently high to assure privacy.
(ii) In all toilet rooms installed on or after August 31, 1971, the floor and sidewalls, including the angle formed by the floor and sidewalls and excluding doorways and entrances, shall be watertight. The sidewalls shall be watertight to a height of at least five inches.
(iii) The floors, walls, ceilings, partitions, and doors of all toilet rooms shall be of a finish that can be easily cleaned. In installations made on or after August 31, 1971, cove bases shall be provided to facilitate cleaning.

(3) Construction and installation of toilet facilities

(i) Every water carriage toilet facility shall be set entirely free and open form all enclosing structures and shall be so installed that the space around the facility can be easily cleaned. This provision does not prohibit the use of wall-hung-type water closets or urinals.

(ii) Every water closet shall have a hinged seat made of substantial material having a nonabsorbent finish. Seats installed or replaced after June 4, 1973, shall be of the open-front type.

(iii) Nonwater carriage toilet facilities and disposal systems shall be in accordance with section 1910.143.

(d) Washing Facilities -

(1) General Washing facilities shall be maintained in a sanitary condition.

(2) Lavatories

(i) Lavatories shall be made available in all places of employment in accordance with the requirements for lavatories as set forth in Table 9.2 of this section. In a multiple use lavatory, 24 linear inches of wash sink or 20 inches of a circular basin, when provided with water outlets for each space, shall be considered equivalent to one lavatory. The requirements of this subdivision do not apply to mobile crews or to normally unattended work locations if employees working at these locations have transportation readily available to nearby washing facilities which meet the other requirements of this paragraph.

(ii) Each lavatory shall be provided with hot and cold running water, or tepid running water.

(iii) Hand soap or similar cleaning agents shall be provided.

(iv) Individual hand towels or sections thereof, of cloth or paper, warm air blowers or clean individual sections of continuous cloth toweling, convenient to the lavatories shall be provided.

(v) Receptacles shall be provided for disposal of used towels.

(vi) Warm air blowers shall provide air at not less than 90°F and shall have means to automatically prevent the discharge of air exceeding 140°F.

(vii) Electrical components of warm air blowers shall meet the requirements of subpart S of this part.

TABLE 9.2 - Minimum Number Of Lavatories

Type of employment	Number of employees	Minimum number of Lavatories
Nonindustrial - office buildings, public buildings, and similar establishments	1-15	1
	16-35	2
	36-60	3
	61-90	4
	91-125	5
	Over 125	Additional fixture for each additional 45 employees
Industrial factories, warehouses, loft buildings, and similar establishments	1-10	1 fixture for each 10 employees
	Over 100	1 fixture for each 15 additional employees

(3) Showers

(i) Whenever showers are required by a particular standard, the showers shall be provided in accordance with subdivision (ii) through (v) of this subparagraph.

(ii) One shower shall be provided for each 10 employees of each sex, or numerical fraction thereof, who are required to shower during the same shift.

(iii) Body soap or other appropriate cleansing agents convenient to the showers shall be provided as specified in paragraph (d)(2)(iii) of this section.

(vi) Showers shall be provided with hot and cold water feeding a common discharge line.

(v) Employees who use showers shall be provided with individual clean towels.

(e) Change Rooms - Whenever employees are required by a particular standard to wear protective clothing because of the possibility of contamination with toxic materials, change rooms equipped with storage facilities for street clothes and separate storage facilities for the protective clothing shall be provided.

(f) Clothes Drying Facilities - Where working clothes are provided by the employer and become wet or are washed between shifts, provision shall be made to insure that such clothing is dry before reuse.

(g) Consumption of Food and Beverages on the Premises -

(1) Application – This paragraph shall apply only where employees are permitted to consume food or beverages, or both, on the premises.

(2) Eating and drinking areas - No employee shall be allowed to consume food or beverages in a toilet room nor in any area exposed to a toxic material.

(3) Waste Disposal Containers - Receptacles constructed of smooth, corrosion-resistant, easily cleanable, or disposable materials shall be provided and used for the disposal of waste food. The number, size, and location of such receptacles shall encourage their use and not result in overfilling. They shall be emptied not less frequently than once each working day, unless unused, and shall be maintained in a clean and sanitary condition. Receptacles shall be provided with a solid tight-fitting cover unless sanitary conditions can be maintained without use of a cover.

(4) Sanitary storage – No food or beverages shall be stored in toilet rooms or in an area exposed to a toxic material.

(h) Food Handling - All employee food service facilities and operations shall be carried out in accordance with sound hygienic principles. In all places of employment where all or part of the food service is provided, the food dispenser shall be wholesome, free from spoilage, and shall be processed, prepared, handled, and stored in such a manner as to be protected against contamination.

(Sec. 6.84 Stat. 1593; 29 U.S.C. 655; CFR 1910.4)
Signed at Washington, D.C., this 26th day of April 1973.

JOHN H. STENDER
Assistant Secretary of Labor

CHAPTER 10

FOOD CONTAINERS & PACKAGING MATERIALS

Food product containers are a most important aspect of all processed and packaged foods and this includes the integrity of the closure.

Cleaning of empty cans, jars and other food containers prior to filling and closing is a most essential step in food processing to prevent contamination with any possible foreign or extraneous materials. The types of contamination vary with how the empty containers are handled prior to their use. Examples of contaminants include dust, dirt, bird droppings, pieces of paper, cob webs, sawdust, grease, rodents, insects and/or microorganisms. Any and all possible contaminants should be removed prior to filling with the food products.

In-house containers made from roll stock, semirigid and flexible containers for food should be clean and free from contaminants. The package making equipment must be kept clean and free of contaminants and the filling operation must be kept clean and in sanitary conditions at all times.

Consideration should be given to the following factors for improving the cleaning of rigid containers prior to filling and closure:

1. Water is the most effective aid in cleaning empty rigid containers when compared to air blast and steam under pressure. However, combinations of water and air or water and steam may be beneficial in some instances.
2. Water pressure should be, at least, 12 psi, but no beneficial effect is gained with water pressure in excess of 15 psi in cleaning efficiency of rigid containers.
3. Water temperature should be, at least, 180° F. (82° C.) for maximum cleaning efficiency of rigid containers.

4. The volume of water will vary with nozzle type and pressure, however, a minimum of 2.5 gallons per minute for washing up to 100 containers per minute has been found to be adequate.

5. The containers should be inserted during the washing cycle for maximum cleaning. All containers must be thoroughly drained prior to filling with food.

6. Spray nozzles designed to give a full cone spray pattern (see figure 10.1) to cover the sides and bottom of the container are more efficient than pipes drilled with holes. The particle size of the nozzle is determined by the orifice opening and the pressure of the water. The orifice opening should be 1/8th to 1/4 inch in diameter.

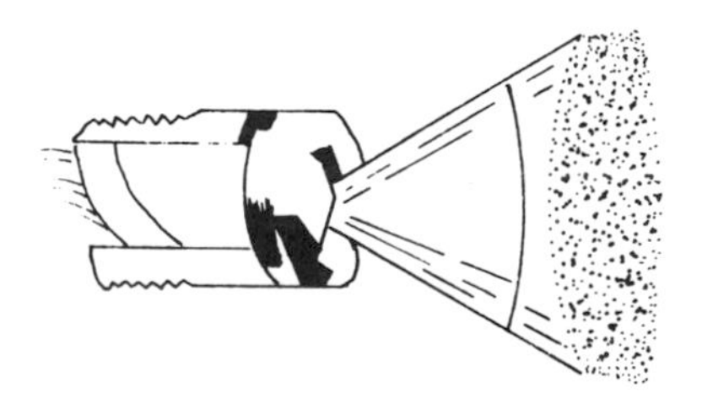

FIGURE 10.1 - FullJet Nozzle Square Spray

7. If air blast or vacuum cleaning are used, the containers should be inverted and it should proceed the washing of the container with water.

8. Non-toxic, non-corrosive properly formulated detergents may be important and useful if the containers should contain oily films.

9. Microbial reductions in containers is enhanced when using water at 180° F. under high pressure.

10. The slower the speed through container washers, the more effective the cleaning of the containers.

FIGURE 10.2 - Can Washing Equipment

The filling and closing of rigid containers requires care to insure the product free of contamination. Overfilling is a major cause of product contamination during closure and processing. Most importantly, great care must be exercised in handling of empty containers prior to filling, during filling and closure, and during the cooking and cooling operations. The total container handling system must be kept clean and sanitized on a regular routine basis to avoid spoilage due to recontamination.

Inspection of containers by qualified personnel at intervals of sufficient frequency to ensure the integrity of the closure is required by FDA (see CFR 113.60 below). In addition, regular observations shall be made during production runs for gross closure defects of glass containers. Any such defects shall be recorded along with the corrective action taken. Visual examinations and teardown examinations shall be conducted, on at least one container from each seaming head or closing machine, with sufficient frequency to ensure proper closure by a qualified container closure technician. Additional inspections shall be made at the beginning of production, immediately following a jam or after machine adjustments. Visual examination should not exceed 30 minutes and tear down examinations should not exceed 4 hours. Physical tests should be made at least every 2 hours on continuous production with semirigid and flexible containers. The inspection results, along with any additional information necessary to ensure proper closure, shall be recorded along with any corrective action taken.

FIGURE 10.3 - Roll Seam Diagram

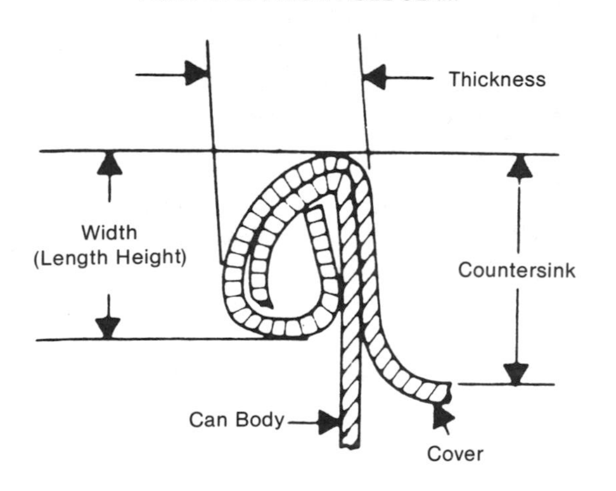

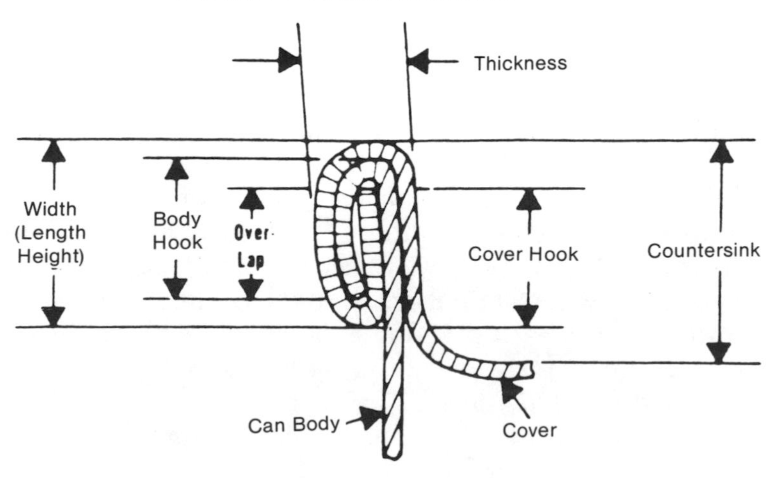

Subpart D - Control of Components, Food Product Containers, Closures, and In-Process Materials

§ 113.60 Containers

(a) *Closures.* Regular observations shall be maintained during production runs for gross closure defects. Any such defects shall be recorded and corrective action taken and recorded. At intervals of sufficient frequency to ensure proper closure, the operator, closure supervisor, or other qualified container closure inspection person shall visually examine either the top seam of a can randomly selected from each seaming head or the closure of any other type of container being used and shall record the observations made. For double-seam cans, each can should be examined for cutover or sharpness, skidding or deadheading, false seam, droop at the crossover or lap, and condition of inside of countersink wall for evidence of broken chuck. Such measurements and recordings should be made at intervals not to exceed 30 minutes. Additional visual closure inspections shall be made immediately following a jam in a closing machine, after closing machine adjustment, or after startup of a machine following a prolonged shutdown. All pertinent observations shall be recorded. When irregularities are found, the corrective action shall be recorded.

(1) Teardown examinations for double-seam cans shall be performed by a qualified individual and the results therefrom shall be recorded at intervals of sufficient frequency on enough containers from each seaming station to ensure maintenance of seam integrity. Such examinations and recordings should be made at intervals not to exceed 4 hours. The results of the teardown examinations shall be recorded and the corrective action taken, if any, shall be noted.

(i) Required and optional can seam measurements:

Table 10.1 - OPTIONAL CAN SEAM MEASUREMENT

(a) Micrometer measurement system:

Required	Optional
Cover hook	Overlap (by calculation)
Body hook	Countersink
Width (length, height)	
Tightness (observation for wrinkle)	
Thickness	

(b) Seam scope or projector:

Required	Optional
Body hook	Width (length, Height)
Overlap	Cover hook
Tightness (observation for wrinkle)	Countersink
Thickness by micrometer	

(c) Can double seam terminology:
(1) "Crossover": The portion of a double seam at the lap.
(2) "Cutover": A fracture, sharp bend, or break in the metal at the top of the inside portion of the double seam.
(3) "Deadhead": A seam which is incomplete due to chuck spinning in the countersink.
(4) "Droop": Smooth projection of double seam below bottom of normal seam.
(5) "False seam": A small seam breakdown where the cover hook and the body hook are not overlapped.
(6) "Lap": Two thicknesses of material bonded together.
(ii) Two measurements at different locations, excluding the side seam, shall be made for each double seam characteristic if a seam scope or seam projector is used. When a micrometer is used, three measurements shall be made at points approximately 120° apart, excluding the side seam.
(iii) Overlap length can be calculated by the following formula:

The theoretical overlap length =
CH + BH + T - W, where
CH = cover hook
BH = body hook
T = cover thickness, and
W = seam width (height, length)

(2) For glass containers with vacuum closures, capper efficiency must be checked by a measurement of the cold water vacuum. This shall be done before actual filling operations, and the results shall be recorded.
(3) For closures other than double seams and glass containers, appropriate detailed inspections and tests shall be conducted by qualified personnel at intervals of sufficient frequency to ensure proper closing machine performance and consistently reliable hermetic seal production. Records of such tests shall be maintained.
(b) *Cooling water.* Container cooling water shall be chlorinated or otherwise sanitized as necessary for cooling canals and for recirculated water supplies. There should be a measurable residual of the sanitizer employed at the water discharge point of the container cooler.
(c) *Coding.* Each hermetically sealed container of low-acid processed food shall be marked with an identifying code that shall be permanently visible to the naked eye. When the container does not permit the code to be embossed or inked, the label may be legibly perforated or otherwise marked, if the label is securely affixed to the product container. The required identification shall identify in code the establishment where packed, the product contained therein, the year packed, the day packed, and the period during which packed. The packing period code shall be changed with sufficient frequency to enable ready identification of lots during their sale and distribution. Codes may be changed on the basis of one of the following: intervals of 4 to 5 hours; personnel shift changes; or batches, as long as the containers that constitute the batch do not extend over a period of more than one personnel shift.
(d) *Postprocess handling.* When cans are handled on belt conveyors, the conveyors should be so constructed as to minimize contact by the belt with

the double seam, i.e., cans should not be rolled on the double seam. All worn and frayed belting, can retarders, cushions, etc. should be replaced with new nonporous material. All tracks and belts that come into contact with the can seams should be thoroughly scrubbed and sanitized at intervals of sufficient frequency to avoid product contamination. Automatic equipment used in handling filled containers should be so designed and operated as to preserve the can seam or other container closure integrity.

REFERENCE

Anon. 1988 Canned Foods. Principles of Thermal Process Control, Acidification and Container Closure Evaluation. 5th Edition. The food Processors Institute, Washington, D.C.

Lucas, Loren L. 1967. Effect of Water, Steam, Air, and Detergent Systems for Removal of Residues from Empty Cans. MS thesis, Ohio State University.

CHAPTER 11

CONTROL OF RATS, MICE AND OTHER RODENTS

Rodents should be controlled in and around a food plant for three basic reasons, that is, (1) the economics of lost products consumed by them and the damage to other products causing their destruction, (2) the aesthetics standpoint, that is, rodents and their filth is repugnant, and (3) from the standpoint that rodents are carriers and transmitters of disease. Good sanitation is the first step in the control of rodents. Management of a food firm must insist on elimination of their entrance and prevention of their breeding in and around the facilities. Proper building design and maintenance of same and the prevention of their harborage are primary steps in control of rodents. The use of baits, traps, chemicals, etc. are all secondary control practices.

Rodents may be defined generally as small animals that have one pair of chisel-like incisor teeth which provide them the ability to gnaw. This gnawing is a vital part of their mode of life. The following is their taxonomy:

Kingdom - Animal
- Sub kingdom - Metazoa
  - Phylum - Chordata
    - Sub phylum - Craniata (Vertibrata)
      - Class - Mamalia
        - Order - Rodentia

Rodentia is the largest order (quantitatively) of mammals and includes squirrels, chipmunks, woodchucks, gophers, beavers, rats, mice, muskrats, lemmings, voles, guinea pigs, capybaras, and chinchillas. From the order, sub-classification begins to branch extensively. There are three sub-orders of Rodentia:

Sciuromorpha - squirrels, beavers, and allies
Hystricomorpha - porcupines and relatives
Myomorpha - rats, mice, voles, etc.

Rats, mice, and chipmunks live on the ground, gophers and woodchucks burrow, squirrels and porcupines climb trees, and muskrats and beavers are semi-aquatic.

Rodents related to food processing are commonly referred to as vermin. Vermin may be defined as any noxious, mischievous or disgusting animal. Food processing vermin are small in size, commonly occur, and generally are difficult to control. Due to the nature of rats and mice and their ability to adapt to new environments, they are of much greater importance in a sanitation program in most food plants than all other rodents combined.

Rats and mice are of the sub-order Myomorpha, the family Cricetidae. They belong to the general Rattus and Mus. The three most destructive and dangerous rodents are: *Rattus Norvegious* (Norway rat, brown rat, barn rat, sewer rat, and grey rat), *Rattus rattus* (roof rat and the black rat) and Mus Muscullus (house mouse).

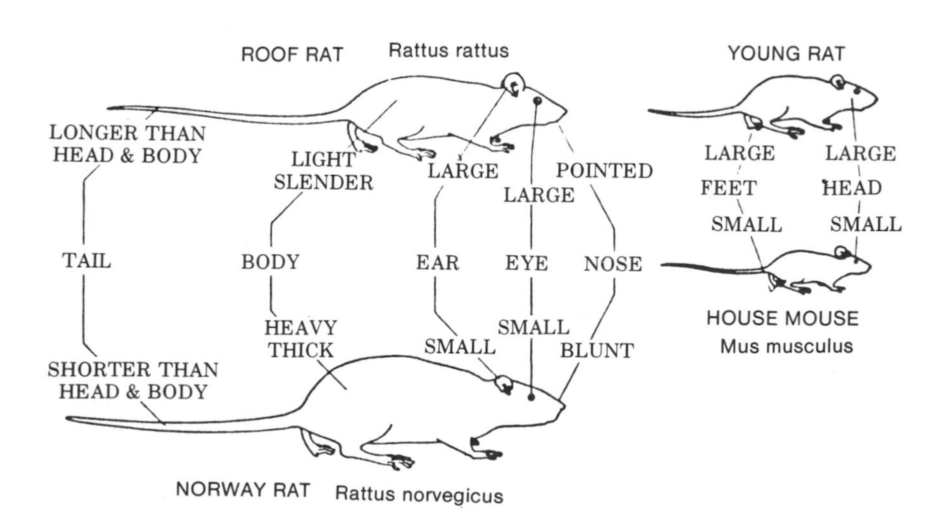

FIGURE 11.1 - Field Identification Of Domestic Rodents
(Courtesy Dept. of Health, Educ. & Welfare, Public Health Service, FDA)

Rattus Norvegious burrows and lives in holes rather than in buildings above ground. The burrows may be 12 to 18 inches below the surface. Further, they are capable of burrowing through hard clay or even lime mortar and sand between bricks or stone walls. They can gnaw holes in lead pipes and slate. They climb and jump poorly and generally are considered to be very clumsy.

Rattus rattus or the black rat prefers to live above ground in hollow spaces in walls, rubbish, attics, or storage areas. It is an agile climber, that is trees, pipes, ropes, and wires.

Mus musculus are excellent climbers and good swimmers and can adapt to various temperatures. They can travel through small openings, that is, ¼ inches to ½ inches in diameter. They live in closer proximity to man than the rat will provided there is suitable hiding places. The mouse is generally dark gray on the back gradually changing to light gray on the abdomen. Variations can occur, all the way from black to brown and an occasional albino. The ears are large and prominent.

The larger rats may grow to 18 inches in length and weigh up to 12 ounces while the mouse will be 5 to 6 inches in length and weigh only about 2 ounces. A large rat can reach up vertically 18 inches and can jump about 2 feet from a still position and up to 3 feet from a running start.

The rats are prolific if food is abundant and they commence to breed when they are 3 months old and they will produce 3 to 8 litters per year. The average litter size is 10. Their life span is 3 to 5 years. The mouse will breed at 2 months of age and will produce about 10 litters per year. The gestation period is 21 days, thus it is possible for a mouse to have 5 to 8 litters averaging 5 young every three weeks. Mice seldom live longer than 2 years.

Rats feed twice during the night - once shortly after dark and again early in the morning. They generally stay within the food plant during their feedings. They may eat up to 1 oz. of dry food and up to 1½ oz. of water every 24 hours.

Mice are primarily seed or cereal eaters, although they will eat the same food as man but prefer foods high in protein and sugar. Mice require little or no water. Mice are nibblers in contrast to rats which, upon locating a suitable food, will simply stay there and eat until they had their fill.

Baiting is generally not effective for mice as they do not eat long enough in one place to obtain enough poison to be killed. Mice will nibble at food during an entire 24 hour period when they are awake. They generally will not travel more than 10-20 feet from their nest.

The most important step in the control of mice and rats is to remove all harborages, including hollow structural and storage areas within the plant. Bait boxes with anticoagulant type baits may be effective for rats. Based on body weight, it takes three days for a fatal dose to be administered to a rat. Bait boxes should be marked on a map of the area treated and they should be checked systematically. Bait boxes must be located at all entrances and around the perimeter of buildings. Generally, they should be set at intervals no less than 30 feet. Obstructions placed in the routes of rats should be used to force the rats through the bait boxes. The routs of rats are generally noted by looking for smears on the walls and floors, their droppings, and evidence of their feeding.

In some cases rats may not take to baiting and trapping must be utilized. This is particularly true of roof rats. The traps should be set directly in a runway with the extended trigger next to a wall and the trap coming out perpendicular to the wall. The trap may be baited with peanut butter or a piece of marshmallow. Traps should be checked systematically with the removal of any dead animals.

To avoid infestation with mice one must make weekly inspections for mice pellets, evidence of their gnawing or the pieces of frass which result from their gnawing. Particular attention should be given to all incoming materials for mice infestation. Spots suspected to be mouse urine can be inspected with an ultraviolet light to determine whether or not that is what it is. Dried urine will fluoresce. However, one should look for pellets as other substances will fluoresce. Mice eat about 4 pounds of food in a six month period and will eliminate about 48 droppings a day. In six months, a pair of mice will eliminate 18,000 droppings and urinate about 1/2 cc of urine in 24 hours. Mice are susceptible to trapping. The traps should be placed in areas where pellets and other evidence of their presence have been found. To bait mice, prepare a mixture of corn meal and bacon fat into a crumbly dough and crumble this across the trap. The mouse will nibble and walk across and, hopefully, get caught. Dead carcasses must be removed daily before they dry out and their hairs fly around the plant. All traps should be numbered and checked by number daily. Box like traps are becoming widely used and very effective as once in the box they will make noise and attract other mice. Again all traps and box traps must be numbered and checked on a daily basis. Accurate records must be kept.

Other rodents may stay around a facility if a food supply attracts them. They can be controlled by trapping them or by exterminating them by poisoning them or by shooting them.

Dogs and cats, also, may be a problem around a food plant, but here again, if the food supply is cleaned up they generally will not remain. If they are persistent they may have to be trapped and turned over to the pound.

Some firms employ outside exterminators for their rodent control program. The plant management is responsible for their activities and should know where the traps and bait boxes are set and what bait is being used. Further, they should have access to the daily records. In most cases the sanitarian or a person he or she supervises can handle any rodent problem. They should be certified as having had training and being competent.

There is no excuse for a rodent problem in a food plant. It is generally only a matter of making appropriate inspections and following through with an action plan. Rodents can be controlled and it behooves every food firm to control them rather than let them they have access to the plant. The best control is to build them out and then follow through with daily inspections of all incoming materials and the premises. This is the only successful way to control rodents.

CHAPTER 12

INSECTS AND INSECT CONTROL

There are more than 700,000 known species of insects. Each of these thousands of species number into billions of individual insects. Some species are large enough to be seen while others are too small or seldom seen.

Insect control in a food establishment is of vital significance in the production of clean and wholesome products. Economic losses plus infested with resulting consumer complaints may be

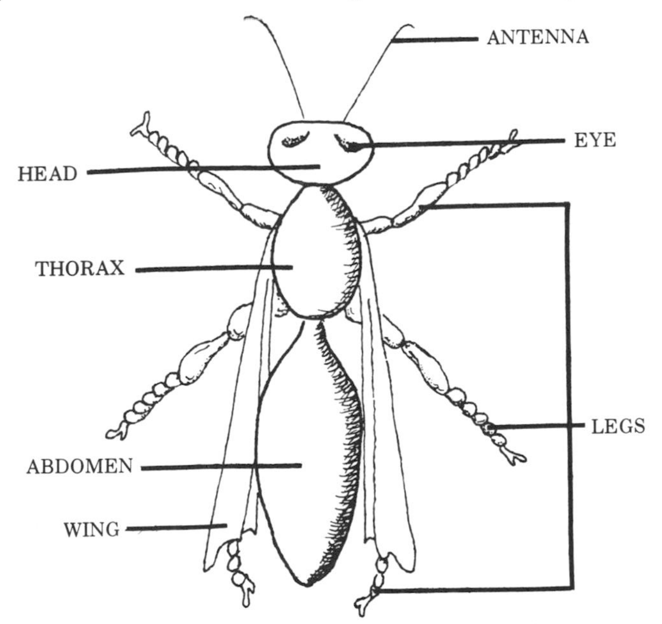

Figure 12.1 - PARTS OF INSECTS

The insect's small body is flexible and well-armored. This has much to do with its success in staying alive. A protective hard shell, lighter in weight than bone, covers the many segments that form three main body regions: head, thorax, and abdomen. Other parts are antennae and eyes found on the head; wings and three pairs of jointed legs attached to the thorax.

serious problems when insects are not controlled. As indicated in Chapter 2, the findings of insect parts is an index to poor sanitation and may be cause for food product seizure and destruction. A diagram of the parts of an insect is shown in Figure 12.1.

Insects do not originate spontaneously. There are several steps to their life cycle is indicated in Figure 12.2. Beetles and moths pass through three stages of development before becoming adults. The adults lay eggs which after hatching become small worms or larvae. After extensive feeding, the larvae develop in size, forming a pupa. The pupa does not feed or move about. In their resting stage, the insect transforms from the worm to an image of an adult or mature insect.

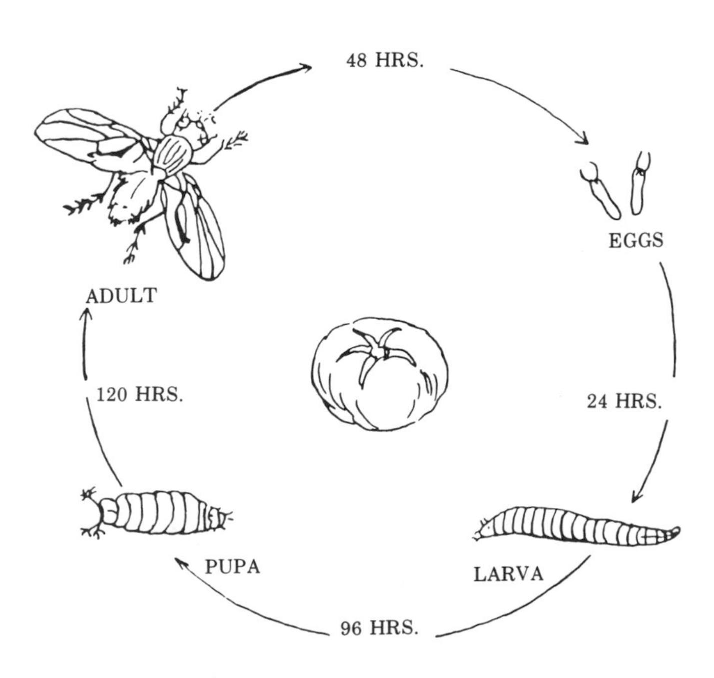

Figure 12.2 - Life Cycle of the Drosophila (Redrawn from Dorst & Knowlton (5))

After a period of time, the pupa splits open and the adult crawls forth. Few species complete the cycle in less than a month and some require as long as a year. Some insects do not have a larva or worm stage in their development but develop from the egg and emerge as dwarf replicas of the adult (see figure 12.3). As this dwarf or nymph feeds, it outgrows its skin. The old skin splits and out crawls a larger form. This shedding occurs several times in the period from egg to adult. Examples of this group of insects are the cockroach and the silverfish.

If insects are so numerous that they can be seen flying or crawling conspicuously about, the source of the infestation should be located and eliminated by cleaning, fumigating, spraying, or screening them out. The finding of any insect or insect part in a food is considered by the public as an index of poor sanitation.

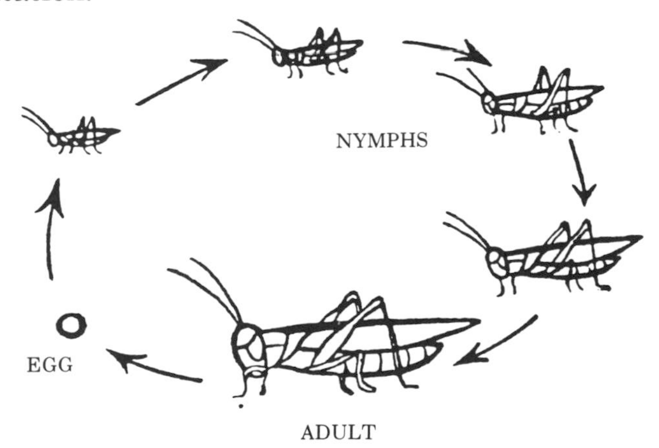

FIGURE 12.3 - LIFE CYCLE OF THE GRASSHOPPER

The grasshopper is an example of gradual growth. The egg hatches, and the tiny insect resembles the adult minus fully developed wings. Through a molting process, the insect sheds its hard skin several times as it grows from the young nymph to the adult.

There are three basic categories of infestation that must be controlled:

Crawling insects--This type is characterized by the different types of cockroaches, silverfish, firebrats, etc. The most common is the cockroach. They are found in every type of food operation and they carry and spread numerous diseases. They are known to carry 4 strains of poliomyelitis, more than 40 different pathogenic organisms and the eggs of several worms. It has been estimated that a single cockroach can carry a total of 13,470 bacteria. Female cockroaches do not lay eggs singly, that is one at a time; they produce small egg cases that contain from 6 to 33 eggs and this egg case is deposited in a hiding place. Young cockroaches begin feeding soon after they hatch. They feed on the same materials as the adults and they look like adults except for size and absence of wings. After shedding their skin several times to grow larger, they become winged adults. Adult cockroaches live for a few months to over a year. They mate several times and the females generally produce one egg case per month. The mouth parts of cockroaches are the biting and chewing type. They can and will feed on a variety of foods, but they prefer starch or carbohydrate type materials. They will sip milk, nibble at cheese, meats, pastry, flour, meal, grease, chocolate, and other foods. They can feed on book bindings, show linings, dead insects, other cockroaches, and human waste. They usually feed at night when they are not likely to be disturbed by human activities.

These cockroach species are common to food processing operations:

German Cockroach - This is probably the most common and widespread cockroach in food plants around the world. It is about 3/4 inch in length, yellowish brown with 2 dark-brown stripes behind the head. Both male and female have well developed wings. The female carries the egg case protruding from the tip of the abdomen until hatching time. The adult female may live for about nine months and produce about 140 young. The egg cases are hidden in areas with abundant food, water, and hiding places. In food processing plants, the German cockroach infests the main production areas, storage areas, offices, lockers and rest rooms. They are not usually found below ground level.

American Cockroach - This is the largest cockroach with adults reaching 2 inches in length. The adult cockroach is brown while the young are pale brown. The female hides her egg cases as soon as they are produced. She may live up to 18 months and lay as many as 33 egg cases or 430 young. The American cockroach usually inhabits basements, storage rooms, garbage areas, and sewers. They are frequently seen in below ground storage areas, and loading docks.

Oriental Cockroach - This insect is about 1 inch long, dark brown to black in color and the wings are very short in the male and absent in the female. The young are pale brown in color. The female hides her egg case as soon as it is formed and she may produce an egg case every month for 5 to 6 months of her life, thus producing some 80 young. The Oriental cockroach likes below ground areas similar to the American cockroach.

Other crawling insects are the Carpet Beetle and the Cigarette Beetle which get into the structure of buildings and feed upon dust and debris. They are generally found between floors and in the ceilings of plants or in cracks in the floors and walls.

Control of crawling insects is by elimination of their harborage. This is done by filling all cracks and crevices and maintaining a sealed smooth surface throughout the plant. The only insecticides permitted in a food plant are those approved by EPA. If they are used, they must be used by a licensed and trained operator and under the strictest of conditions. All insecticides are adulterants if they should get into the finished product. Management is held responsible in all cases.

Flying insects - The most common flying insects are the house fly and fruit fly. They are generally seasonal in the North, but may be prevalent year around in warm climates.

Housefly - This insect is found all over the world and they do spread pathogenic organisms. It has been estimated that one fly can carry up to 3,680,000 bacteria. The organisms are collected on the feet and mouth parts of the fly as they visit garbage and food. The organisms are deposited when the fly crawls on human food or they are deposited in fly increment. The house fly passes though three stages on its way to becoming an adult. The egg hatches in 8 to 12 hours and the maggot that hatches from the egg begins feeding and growing immediately. After 5 days the maggot changes to the pupa stage and rests for about 4 days. The adult fly then emerges and starts the cycle all over again. The adult fly enters buildings in search of food and shelter. Control must be aimed at preventing their entrance by proper screening and once inside by reducing their numbers. Electric fly traps, operating day and night, are one of the best methods for reducing their numbers. Insect-O-Cutor offers three systems: a wall unit, a hanging unit and a portable unit. It uses near-ultraviolet light to attract the flies, which fly between the electric grids. They are electrocuted and then fall into a removable tray.

FIGURE 12.4 - House Fly

Fruit flies - These pests are about 1/10 inch in length with light brown bodies and red eyes. The flies are attracted to fruit and animal wastes, particularly decaying products. Their life cycle is shown in Figure 12.2. Control of the fruit fly is by reducing the amount of waste or rotten products in and around the plant, proper screening (16 mesh), and the effective use of electric fly traps.

Stored Food Insects - These are characterized by the types of insects inherent in certain types of stored foods (see Figure 12.4). This group is principally represented by the Confused Flour Beetle and the Saw-Tooth Grain Beetle which may come into the plant or warehouse in the egg form or as microscopic sized larva already present in the food. Control is usually by fumigation.

Other insects and their characteristics are shown in Table 12.1. Control of all insects can be summarized as follows:

1. Eliminate their harborage places,
2. Eliminate their food and water,
3. Maintain sound structures with appropriate screenings, and
4. Use pesticides only under the supervision of a licensed operator.

REFERENCES

Anon. 1952 Insects, the Yearbook of Agriculture. U. S. Government Printing Office.
Anon. 1956 Handbook of the Insect World. Agr. Chem. Div. Hercules Powder Company, Wilmington, Delaware
Gould, Wilbur A. 1980. Good Manufacturing Practices for Snack Food Manufacturers. Snack Food Assn., Alexandria, VA
Gould, Wilbur A., Geisman, J. R. and Sleesman, J. P. 1959. Washing Tomatoes. OARDC Res. Bull. 825, Ohio Agricultural Experiment Station, Wooster Ohio
Metcalf, C. L. and Flint, W. P. Destructive and Useful Insects. McGraw Hill Book Co., Inc.
Robinson, W. H. 1978. Food Pests and Their Control, and the Use of Pesticides in Food Processing Plants. Sanitation Notebook for the Seafood Industry. Dept. of Entomology, Va. Polytechnic Institute and State University. Blacksburg, VA.
Schoenherr, W. H., A Guide to Brewery Sanitation. Lauhoff Grain Company

FIGURE 12.5 - Principal Stored Grain Insects

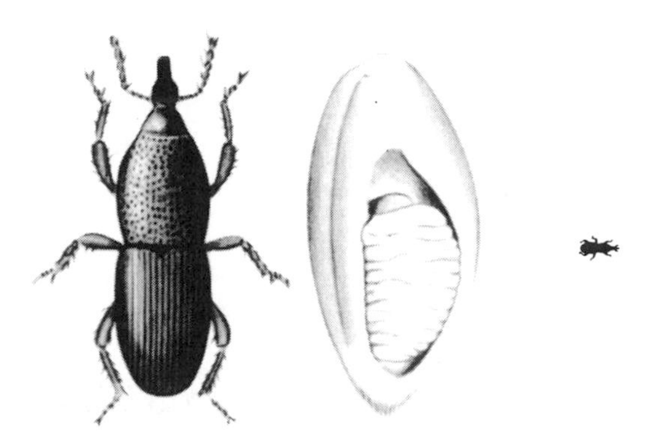

GRANARY WEEVIL, *Sitophilus granarius* (Linnaeus). This true weevil, along with the closely related rice weevil, is among the most destructive of all stored grain insects. The larvae develop inside kernels of whole grain in storage, thus making an infestation difficult to remove in the milling process. In Indiana, the granary weevil is largely a pest of stored wheat, corn and barley, especially in elevators, mills and bulk storages. The adult cannot fly, and field infestations do not occur.

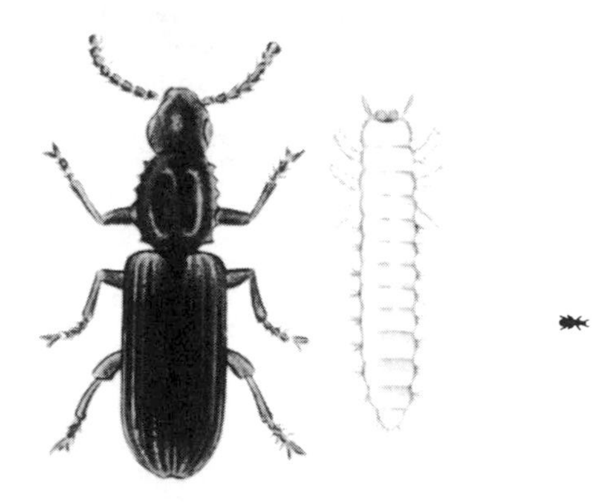

SAW-TOOTHED GRAIN BEETLE, *Oryzaephilus surinamenis* (Linnaeus). Along with flour beetles, the saw-toothed grain beetle is one of the most common insects in stored grain and cereal products. The larvae develop in flour, cereal products and many other dried foods. For this reason, it is a common pest not only in grain bins, but also in elevators, mills, processing plants, warehouses and kitchens. In grain bins, it feeds on broken kernels and grain residues.

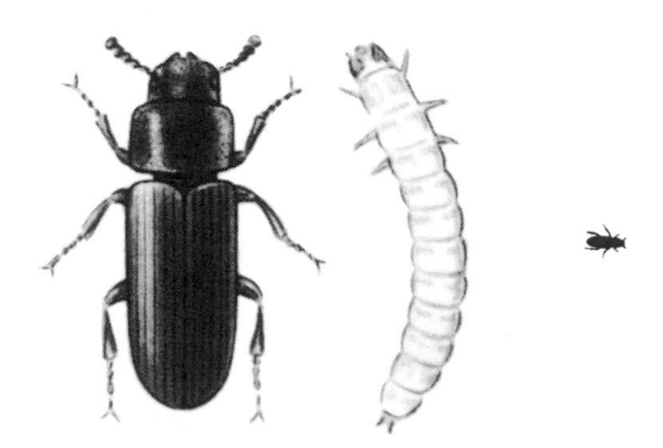

RED FLOUR BEETLE, *Tribolium castanenum* (Herbst). This beetle is similar to the saw-toothed grain beetle in habits and types of products infested. It is a serious pest in flour mills and wherever cereal products and other dried foods are processed or stored. Like the confused flour beetle (not pictured), the red flour beetle may impart a bad odor that affects the taste of infested products.

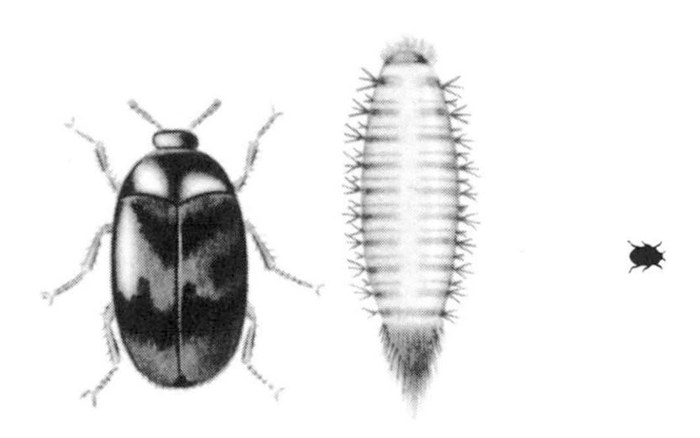

LARGER CABINET BEETLE, *Trogoderma inclusum* (LeConte). Representing a group also referred to as Trogoderma, the larger cabinet beetle is a scavenger that feeds on cereal products and dried animal matter. The fuzzy, slow-moving larvae similar to the larvae of carpet, hide and larder beetles are often found crawling about on or near the products they infest.

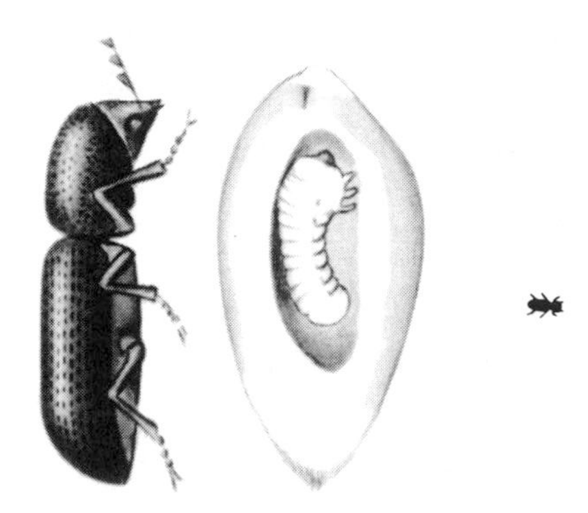

LESSER GRAIN BORER, *Rbyzopertba dominica* (Fabricius). This pest is most common and destructive in warm climates but can spread to any area in transported grain. It is a problem of grain only and not cereal products. The larvae develop inside the kernels of whole grain. The adults also damage grain by boring into the kernels and leaving them covered with powder from the chewed material.

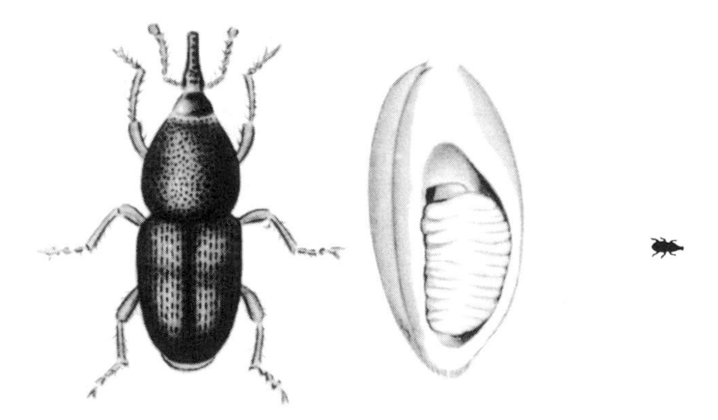

RICE WEEVIL, *Sitopbilus oryzae* (Linnaeus). The rice weevil is similar to the granary weevil in both appearance and habits. The name is misleading, however, since it infests other grains besides rice. Adults can fly and, in warm climates, can cause widespread damage to corn, wheat and other grains before harvest. Although field infestations do not occur in Indiana, post-harvest infestations do. Such infestations originate from shipped-in grain or from already-infested storages.

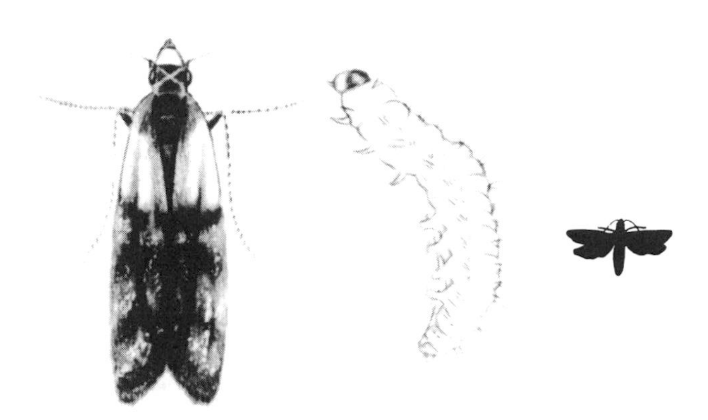

INDIAN-MEAL MOTH, *Plodia interpunctella* (Hubner). Common to both stored grain and cereal products, Indian-meal moth larvae cause damage in corn meal, packaged foods, bagged grain and grain in storage. Attack is confined to surface layers of stored shelled corn and small grains. In the case of stored ear corn, however, feeding occurs anywhere, since the moths crawl among the ears to lay their eggs. Larval feeding is characterized by a webbing of the material infested. The mature larvae then often leave the material and crawl about in homes or buildings in search of a place to pupate.

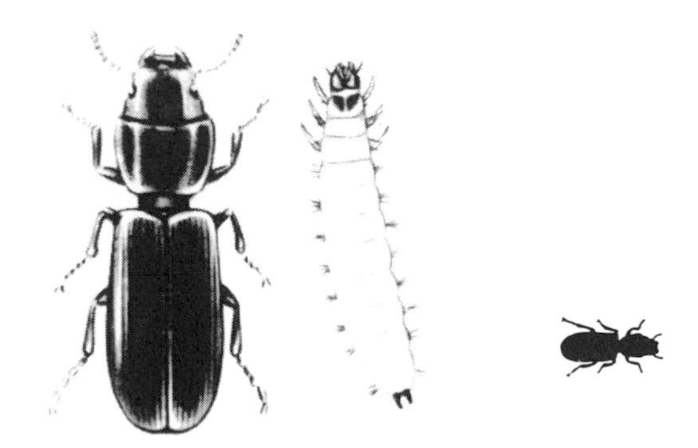

CADELLE, *Tenebroides mauritanicus* (Linnaeus). Both the adult and larva are large and easy to see. Both stages feed mainly on the germ of stored grains, but may also attack milled cereal products. The larvae leave stored grain in the fall and burrow into woodwork, such as wooden bins or boxcars, to hibernate. They may also burrow into packaged cereal products, thus providing an entrance for other cereal pests.

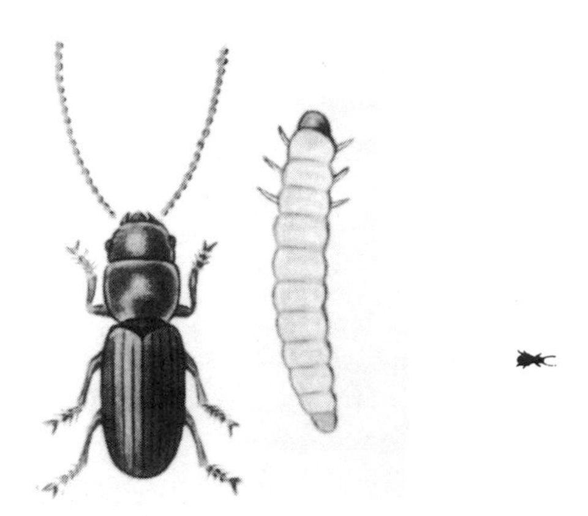

FLAT GRAIN BEETLE, *Cryptolestes pusillus* (Schonherr). This is a tiny beetle that feeds primarily on the germ of stored grains, especially wheat. It is readily attracted to high-moisture grain. In fact, under high-moisture conditions, the flat grain beetle may also develop in many cereal products, but it is not a common pest in kitchens.

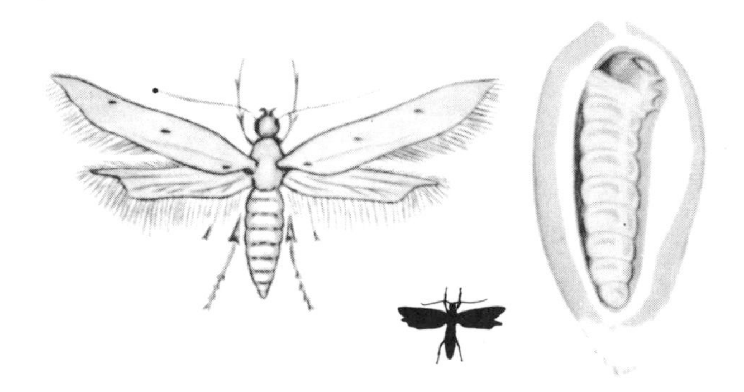

ANGOUMOIS GRAIN MOTH, *Sitotroga cerealella* (Olivier). This is a common and destructive pest of crib ear corn. It also infests stored shelled corn and other small grains, but attack is confined to the surface layer of grain. Field infestations are common in the southern half of Indiana. The larvae develop within the kernels; therefore, the Angoumois grain moth is not a pest of cereal products. Infestations in homes often occur in stored popcorn or in colored ears of corn kept for decoration purposes. The moth resembles the clothes moth but does not shun light.

FIGURE 12.6 - Identification Of Worm And Insect Fragments

APHID

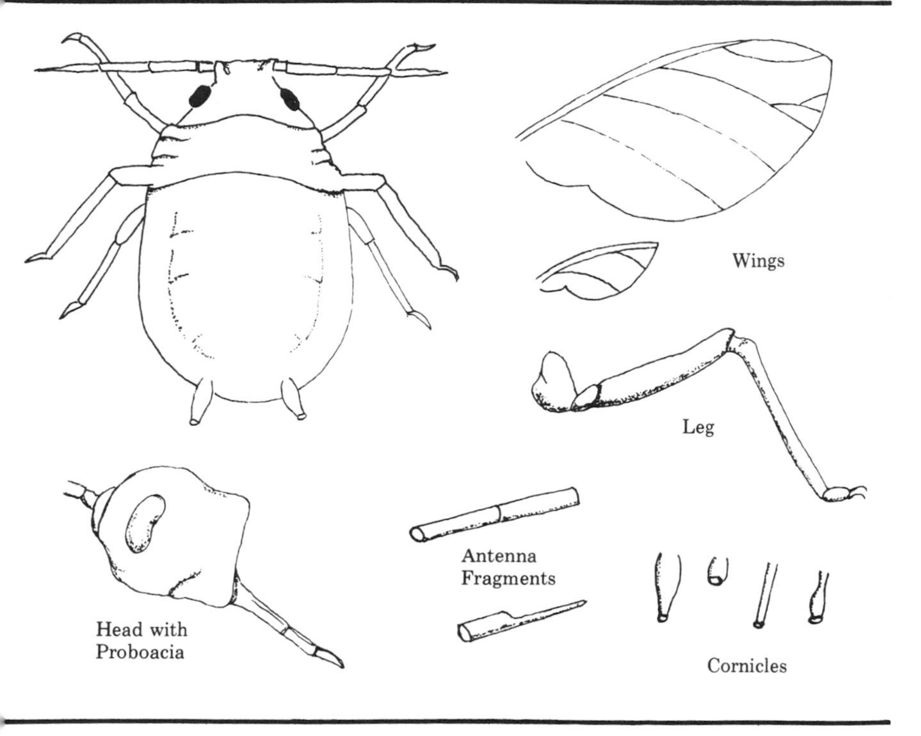

VINEGAR FLY

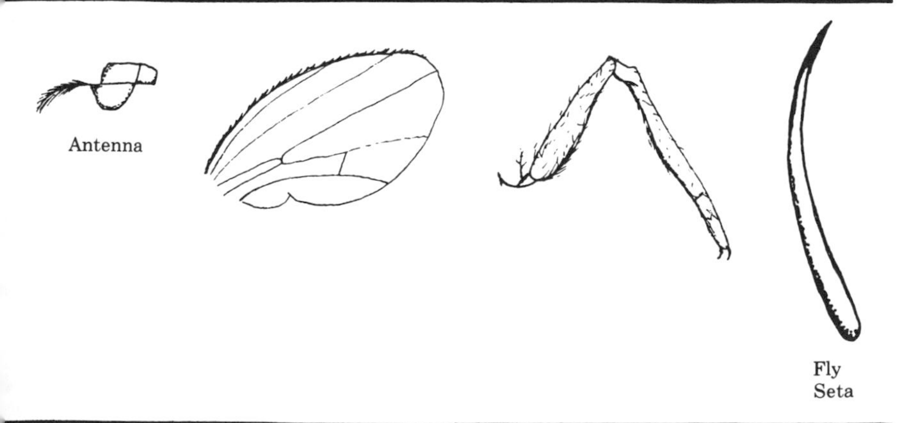

MOTH LARVA (WORM)

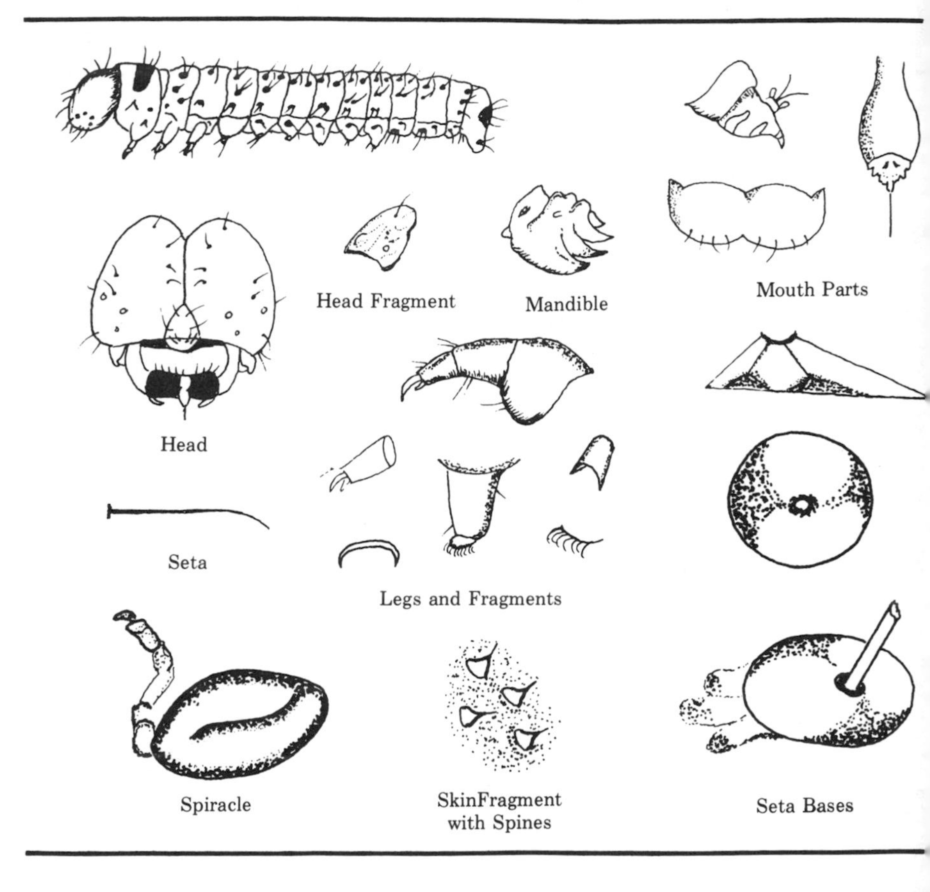

HOUSE FLY

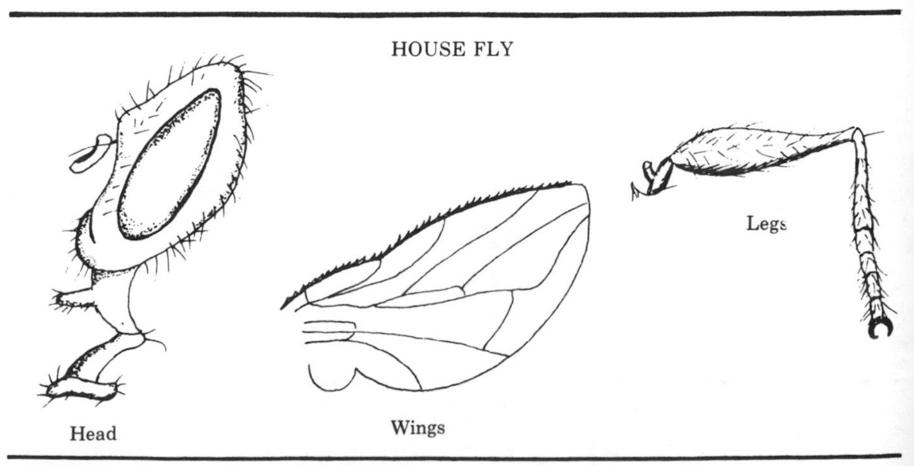

THRIPS

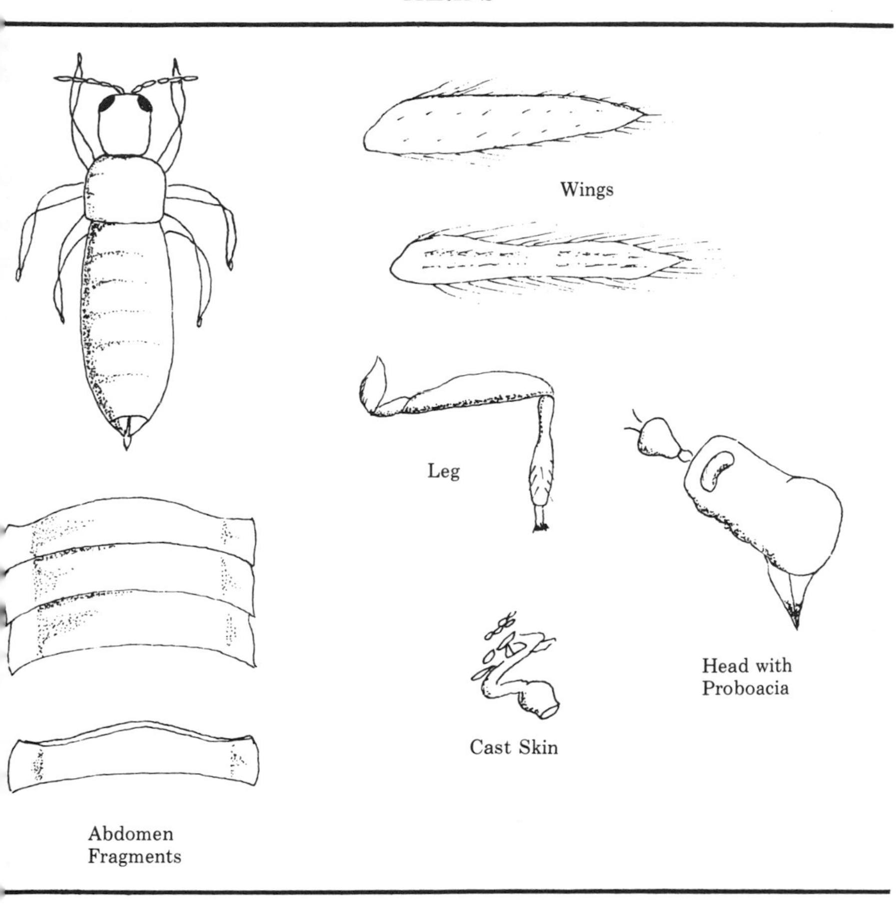

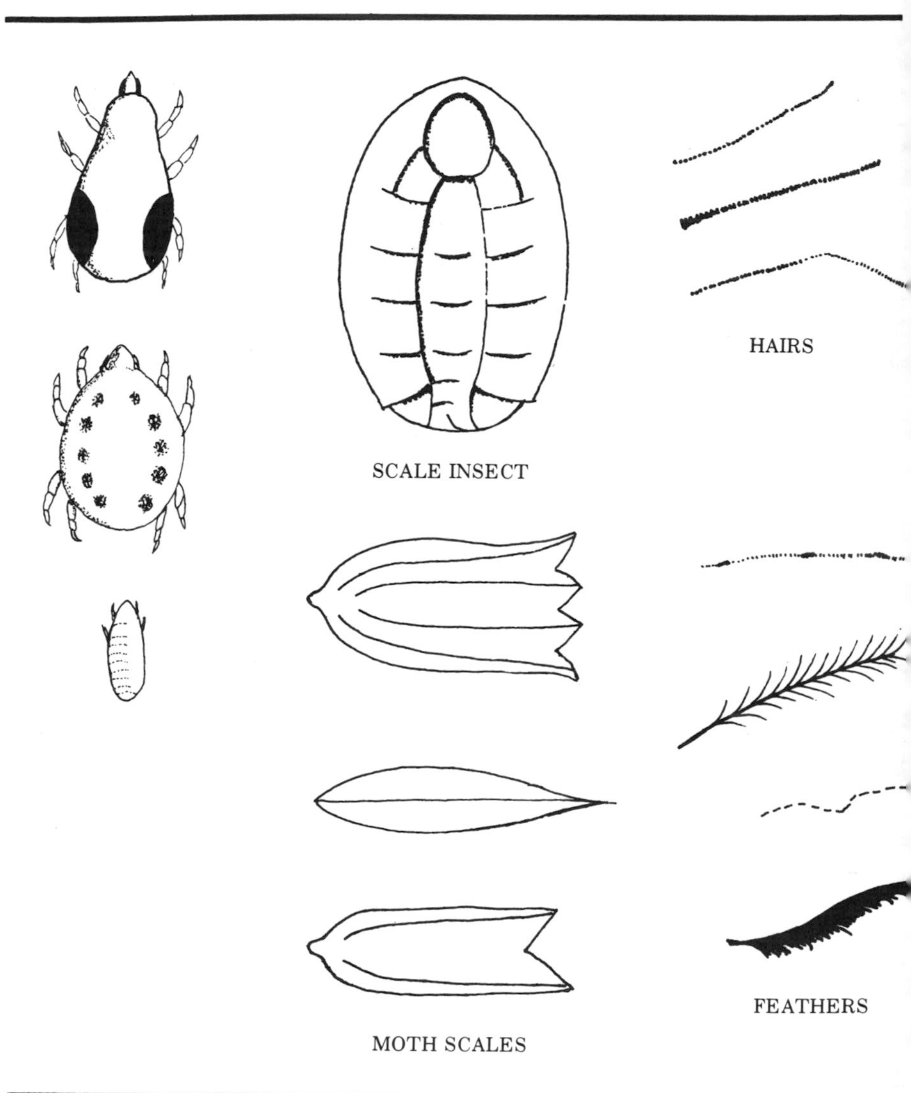
MITES
SCALE INSECT
HAIRS
MOTH SCALES
FEATHERS

TABLE 12.1 - Common Insects Infesting Food, Dairy And Beverage Plants

Common Name	Life Stage	Life Cycle and No. Eggs Laid	Average Reproduction Rate Per Year	Identification and Characteristics
ANT, (PHARAOH) (Monomorium pharaonis)	Egg Larva Pupa Adult	60 days •••• Many thousand eggs by queen ant.	3,000	Black — 1/16" length. Found in all food plants n‹ sweets. Live in colonies. Adulterates products.
BEETLE, CADELLE (Tenebroides mauritanieus)	Egg Larva Pupa Adult	3 months •••• 1,000 eggs	2,000	Black — 3/8" length. Larva bores in wood enabling other insects to hide in burrows. Found in rice and flour mills, candy factories and bakeries.
BEETLE, CARPET (Anthrenus scrophulariae) (Similar to Cigarette, Drug-Store and Larder)	Egg Larva Pupa Adult	3 months to 1 year •••• 50 eggs	500	Black or marbles covered with scales — 3/16" long. Adults can fly; often found on window sills and on flowers. Casts its skin from 5 to 6 times in the larval stage during growth. Nicknamed "Buffalo Moth". Found in cracks in floors, dried milk and feeds upon carpets in buildings. Perforates packages, letting other insects in.
BEETLE, CIGARETTE (Lasioderma serricorne) (Similar to Carpet, Drug-Store and Larder)	Egg Larva Pupa Adult	60 days •••• 100 eggs	5,000	Brown — 1/8" length. Head bent downward. Adults fly. Found in tobacco warehouses, flour mills, bakeries and cotton-seed meal. Adulterates food and perforates packages.
BEETLE, CONFUSED FLOUR (Tribolium confusum)	Egg Larva Pupa Adult	4 weeks •••• 400 eggs	3,000	Brown — 1/3" in length. Adults do not fly. Worst pest in flour mills and bakeries. Found in all food and beverage plants, flour mills and seed companies.
BEETLE, DRUGSTORE (Stegobium paniceum) (Similar to Carpet, Cigarette and Larder	Egg Larva Pupa Adult	6 weeks •••• 100 eggs	3,000	Brown — 1/8" length. Adults fly. Can thrive on poisons such as Strychnine. Found in candy factories and spice and flour mills. Destroys food by perforating packages and letting other insects in.
BEETLE, FLAT GRAIN (Laemophloeus minutus)	Egg Larva Pupa Adult	6 weeks •••• 200 eggs	500	Reddish Brown — 1/16" length. Smallest beetle infesting grain. Found in flour mills and bakeries. Attacks the germ in kernels of wheat.
BEETLE, LARDER (Dermestes lardarius) (Similar to Carpet, Cigarette and Drug-Store	Egg Larva Pupa Adult	60 days •••• 150 eggs	1,000	Dark brown with yellow band across back — 5/16" long. Adults also found outdoors hiding in barks of trees; larva casts its skin 4 or 5 times during growth. Found where food such as ham, bacon, and dried meats are stored. Destroys dried meats, feathers and hair.
BEETLE, SAW-TOOTHED GRAIN (Oryzaephilus surinamensis)	Egg Larva Pupa Adult	30 days •••• 150 eggs	3,000	Dark brown — 1/10" length. Has 6 spines on thorax. Looks like a brown ant. Found in bakeries, seed companies and flour mills. Ruins seeds so they will not germinate.
BEETLE, WHITE-MARKED SPIDER (Ptinus fur)	Egg Larva Pupa Adult	1 year •••• 100 eggs	50	Reddish brown with four white marks on back — 1/8" long. Has the appearance of a tiny spider; walks steadily and slowly. Found in flour mills, bakeries and seed companies. Destroys foodstuffs, spices, cotton bags and seeds.

TABLE 12.1 - Common Insects Infesting Food, Dairy and Beverage Plants - Continued

Common Name	Life Stage	Life Cycle and No. Eggs Laid	Average Reproduction Rate Per Year	Identification and Characteristics
BOCKLOUSE (Liposcelis divinatorius)	Egg Nymph Adult	150 days •••• 60 eggs	500	Grayish — 1/25" length. Wingless. Lives on microscopic molds. Found in flour mills and baking companies. Adulterates products.
CENTIPEDE, HOUSE (Scutigera forceps)	Egg Adult (Egg hatches within body)	1 to 2 years •••• None laid; young born alive	10	Gray in color — 1" length; has 15 pairs of legs — last two pairs longest. Not a true insect. Feeds on insects. Found in damp basement and boiler rooms in every industry and institution.
COCKROACH, AMERICAN (Female) (Periplaneta Americana)	Egg Nymph Adult	2 to 3 years •••• 14 eggs per egg-case; 50 egg cases.	20	Dark brown — 1-1/2" in length. Female carries egg-case. Only of three roaches which fly; chewing mouth-parts. Most powerful of the three roaches; casts its skin 12 times to complete growth. Found in damp basement, boiler rooms and elevator shafts in every industry and institution. Destroys foodstuffs, labels and cartons. Carries filth.
COCKROACH, GERMAN (Female) (Blattella germanica)	Egg Nymph Adult	250 to 300 days •••• 30 to 40 eggs per egg-case; 2 to 5 egg-cases	4,500	Light brown — 5/8" length; female carries egg-case; chewing mouth-parts. Shuns light; runs rapidly; casts its skin 6 times to complete growth. Found in washrooms and lunchrooms in every industry and institution. Leaves odors; destroys packaged goods.
COCKROACH, ORIENTAL (Female)	Egg Nymph Adult	500 to 800 days •••• 13 eggs per egg-case; 9 egg cases	100	Pitch black — 1" males have wings; female wingless and carry egg-case; chewing mouth-parts. More social in habits than other roaches; casts its skin 10 times to complete growth. Found in damp basements, boiler rooms and elevator shafts in every industry and institution. Destroys foodstuffs; carries filth.
CRICKET, HOUSE (Gryllus domesticus)	Egg Nymph Adult	18 months •••• 40 to 170 eggs	10	Black — 1" length. Jumps similar to common grasshopper. Produces chirruping sounds; related to roaches. Casts its skin 9 to 11 times to complete growth. Found in damp basements and elevator shafts in every industry and institution. Feeds on fabrics.
FIREBRAT (Thermobia domestica)	Egg Nymph Adult	3 months to 2½ years •••• 50 eggs	100	Grayish — 1/2" in length. Body covered with scales; chewing mouth-parts. Feeds on paper goods. Found near boiler and steam pipes in every industry and institution. Breeds in stored paper. Destroys paper labels and books.
FLY, DRAIN (Psychoda alternata)	Egg Larva Pupa Adult	8 to 24 days •••• 30 to 100 eggs	1,000	Black — 3/16" wing spread. Moth-like. Not a steady flier; walks on walls near drains. Found in washrooms, usually in sink drains and urinal drains in every industry and institution.
FLY, VINEGAR (Drosophila melanogaster)	Egg Larva Pupa Adult	10 days •••• 50 eggs	5,000	Brown — 1/8" wingspread. Very difficult to see; can breed in flour paste and pickle brine. Found in dairies, bottling plants, breweries, bakeries, candy factories and canneries. Adulterates products.
HOUSEFLY (Musa domestica)	Egg Larva Pupa Adult	8 days •••• 400 to 600 eggs	50,000	Almost black in color — 3/8" wingspread; two wings and sucking tube. Can have 10 to 12 generations in one summer. Found in every industry and institution. Carries diseases and adulterates products.
MILLIPEDE (Paraiulus venustus)	Egg Young Adult	1 year •••• 50 eggs	30	Brown — attains length of 1". Cylindrical; has two pairs of very short legs on each body segment. Normally outdoors. Occasionally enters basements of buildings. Prefers dampness; dryness destroys them. If touched when crawling they curl up. May crawl up walls and fall into processed foods.
MITE, FLOUR (Tyroglyphus farinae)	Egg Hypopus Adult	6 months •••• 100 eggs	300	Microscopic — 10,000 can be placed in square inch of space. Not true insect — has 8 legs; sucking mouth parts. Can lie dormant for many months without food. Found in flour mills, bakeries, candy factories and cheese factories. Adulterates products.

TABLE 12.1 - Common Insects Infesting Food, Dairy and Beverage Plants - Continued

Common Name	Life Stage	Life Cycle and No. Eggs Laid	Average Reproduction Rate Per Year	Identification and Characteristics
MOTH, ALMOND (Ephestia cautella)	Egg Larva Pupa Adult	3 months •••• 100 eggs	500	Gray — 5/8" wingspread. Flies with a quick dart. Very serious pest in raisins, figs and cocoa beans. Found in candy factories, baking and nut companies and chocolate factories. Destroys chocolate nut products and dried fruits.
MOTH, ANGOUMOIS GRAIN (Sitotroga cerealella)	Egg Larva Pupa Adult	5 weeks •••• 40 to 350 eggs	1,000	Light grayish brown — 1/2" wingspread; hind wings bordered with a long delicate fringe. Adult moth lays eggs in growing grain as well as stored grain. Found in flour and feed mills, granaries and railroad box-cars. Also found in field or ripening corn and wheat as they are nearing maturity. Bores holes in corn and wheat making the products unfit for human consumption.
MOTH, INDIAN MEAL (Plodia interpunctella)	Egg Larva Pupa Adult	4 weeks •••• 300 eggs	3,000	Gray and reddish-brown — 1/2" wingspread. Most common moth found on seeds and nuts. Found in candy factories, seed companies, flour mills and nut companies. Destroys foodstuffs by spreading webbing.
MOTH, MEDITERRANEAN FLOUR (Ephestia kuehniella)	Egg Larva Pupa Adult	5 weeks •••• 200 eggs	3,000	Gray with black zig-zag markings — 5/8" wingspread. Most common moth found in flour mills and bakeries. Found in flour mills and baking companies. Destroys foodstuffs by spreading webbing.
PILLBUG (Armadillidium	Egg (hatches within body Nymph Adult	12 months •••• None laid; young born alive	25	Dark brown — 1/2" length. Found in damp basements and elevator shafts in every industry and institution. Rolls up into ball when molested. Adulterates product.
SILVERFISH (Lepisma saccharina)	Egg Nymph Adult	3 months to 3 years •••• 100 eggs	100	Silvery in color — body covered with scales; chewing mouth-parts. Found near boiler and steam pipes in every industry and institution. Feeds on vegetables and animal matter. Destroys labels and cartons.
SKIPPER, CHEESE (Piophila casei)	Egg Larva Pupa Adult	8 days •••• 200 eggs	8,000	Looks like a tiny housefly — has wingspread 3/16"; shiny black. Larva of these flies can hop, or skip, 4 inches at intervals. Found in cheese factories, meat-packing plants and grocery companies. Attacks and damages cheeses and meat products.
WEEVIL, COFFEE-BEAN (Araecerus fasciculatus)	Egg Larva Pupa Adult	60 days •••• 200 eggs	1,000	Dark brown — 3/16" long; has tiny hairs on wing covers. Breeds in kernels of corn in the field and continues to breed after the corn has been placed in storage; also a serious pest in stored green coffee. Found in granaries, seed companies and wholesale grocery warehouses. Destroys cereal products.
WEEVIL, GRANARY Sitophilus) granarius)	Egg Larva Pupa Adult	4 weeks •••• 300 eggs	4,000	Shiny black — 1/8" length. Has long snout. Can produce 6,000 offspring per year. Found in flour mills and baking companies. Bores holes and destroys grain.
WEEVIL, RICE (Sitophilus) oryza)	Egg Larva Pupa Adult	4 weeks •••• 300 eggs	6,000	Black with 4 red spots on back — 1/8" length. Long snout. Plays dead when molested. Found in flour mills, baking companies and macaroni plants. Bores holes in finished macaroni and grain.

CHAPTER 13

CONTROL OF BIRDS

Birds are a nuisance in and around a food establishment. Further, they may contaminate incoming materials, products or materials in process or storage, and outgoing materials. They cause objectionable odors, deface property and they may be the cause of diseases. The most troublesome birds are pigeons, house sparrows, and starlings. Other species may be found.

Birds use large buildings as nesting places in the summer and roost in any season. Birds enter high unscreened areas or openings near the roof. Their droppings are found especially under their roosts. Their droppings are easily recognized by the characteristic appearance of the white chalky matrix. Since most birds are chiefly insectivorous, their excreta often contains undigested fragments of insect cuticle. In addition to their excreta, birds may discard insect fragments in still another way. Many birds (particularly starlings) eat their food voraciously as they find it. After the more easily digested parts have been dissolved, the remainder is discarded through the mouth in the form of castings or regurgitated pellets.

FIGURE 13.1 - Sparrows

Sparrows (Figure 13.1) and pigeons (Figure 13.2) are two of the most common birds found around buildings. The nests of sparrows and pigeons which are often built around the roof of a large building may contaminate food with falling sticks, straw or feathers. An added problem here is that birds that are raised in an area tend always to come back to this area to raise their young, so that there is a constant reinvasion of birds in the nesting areas. The reproduction rates of common bird pests is shown in Table 13.l.

FIGURE 13.2 - Pigeons

Some products may be contaminated in the field with bird excreta or feathers. Many fruits and vegetables, as well as grain, are attacked by birds, and may thus show traces of their presence. This contamination can normally be removed in the preparation of the products for processing.

Usually foods are properly protected but, commercially speaking, there is a feeling of repulsion on the part of customers when they receive products that are covered with bird manure or feces, even in small quantity. Generally this problem is caused by someone leaving doors open and birds enter the warehouse or production sections. They can also enter through ventilation holes or open windows unless screened. Birds can usually be screened out or built out by positive air ventilation. If they do get into the building they must be removed and their nests must be knocked down whether they contain young fledglings or not.

Birds can usually be prevented from entering areas by use of a rotating "bird-light". This is a special designed piece of equipment which has a rotating yellow light that scares birds away when they come back to roost. It is of no value as illumination for a warehouse or room. It must be used in semi-dark areas indoors or in a dark dock area with open rafters.

If the use of the bird light is not applicable and birds are found to roost (if not nest) on areas where they can defecate below and soil the storage or food equipment, then the rafters and roosting areas above can be covered with a liquid bird repellent which aggravates the birds by getting on their feet. They leave and will almost never return.

All birds do not travel in flocks, but certain species such as sparrows and pigeons do, and they will tend to congregate outside the building and wait until there in an opportunity for one or two of them to dart in. They find feeding grounds inside. They are annoying and it is difficult to keep some from getting into the plant.

These can be controlled fairly well by the use of a chemical called AVITROL, which is mixed with their feed. One should scatter the area where they tend to flock with untreated seed for about a week, then for a day or so, set out the treated feed. The birds are led to the area to feed. When they get the treated feed, they are affected and they act silly. They are not killed except in rare instances, but their queer actions are such that the birds become frightened and never come back.

The control of birds is a must for good sanitation and for protecting incoming, in-process and finished products. The first step is to build them out and make certain the facility is well maintained. The second step is to always be on the watch for signs of them and take precautions before they become a problem. They can be controlled and control must be practiced for the good of the firm and his products.

TABLE 13.1 - THE REPRODUCTION RATES OF COMMON BIRDS*

	Sparrow	Pigeon	Starlings
No of young/clutch	5-8	1-2	3-6
No of Broods/yr	3-5	5-6	2
Incubation period(days)	12-13	17-22	12
Average production/yr	25	10	10

* Taken from Parker and Litchfield

REFERENCES

Anon. 1968. General Principles of Food Sanitation, Food and Drug Technical Bulletin No.l. US Dept. of Health, Education, and Welfare. FDA, Washington D.C.

Gould, Wilbur A. 1980. Good Manufacturing Practices for Snack Food Manufacturers. PC/SFA, Alexandria, VA

Parker, M. E. and Litchfield, J. H. 1962. Food Plant Sanitation. Reinhold Publishing Co. Chapman & Hall Ltd. London

CHAPTER 14

FOOD PLANT MICROORGANISMS

Microorganisms are organisms that are invisible or barely visible by the naked eye. They include bacteria, viruses, protozoa, algae, mold, yeast, and certain small parasitic worms. Some microorganisms are useful while others are very harmful. Soils and water are the most common sources of microorganisms. Crops that grow in the soil or near the soil usually carry large numbers of microorganisms.

The smallest common bacteria are only 0.15 um wide and 0.3 um long. Most of the important bacteria in foods are 0.5 to 2.5 um in diameter and 2 to 10 um in length. Yeast cells are much larger ranging from 2.0 to 7 um in diameter. Molds are multicellular and their components vary in size with the mold species. Viruses are the smallest units of living matter and have the simplest structures.

Molds exhibit some of the characteristics of higher plants. They vary greatly in shape, size, color, and reproductive processes. Molds lack chlorophyll and generally prefer dead organic matter or living host tissues for their existence. Molds are composed of vegetative cells or threads (hyphae) called the mycelium. The reproductive structures are called spores. Molds are much larger than bacteria and in some cases larger than yeast. Molds are widely distributed in nature, both in the soil and in dust carried by air. Under suitable conditions of moisture, aeration and temperature, molds will grow on almost any food. The discolored appearance on many moldy foods is due to the fruiting bodies. Molds are ubiquitous and can survive on a wide variety of substances. They may be found growing readily on walls and ceilings if the humidity is high. They can, also, grow in refrigerated storage areas if the temperature is above freezing. Most molds have little heat resistance and are easily killed by thermal processing.

Yeasts are unicellular microscopic living bodies, usually egg-shaped. They are smaller than bacteria and reproduction generally occurs by budding. They are widely found in nature and are generally associated with foods containing high sugars and acids. Most yeast are destroyed by temperatures of 77°C. Yeasts are more tolerant to cold than heat, similar to molds. Figure 14.1 illustrates some types of molds, yeast, and bacteria.

In food processing and preservation it is essential for the sanitarian to know the "critical control points" in a food operation. These are the most likely points where microorganism and other contamination may take place. It's essential that these points be checked and evaluated as part of any quality assurance operation. This area will be discussed in more detail in Chapter 20.

Microorganisms may be classified according to optimal temperature for growth, their peculiar growth requirements, or the type of nutrient they metabolize.

FIGURE 14.1 - The Microorganisms Of Concern In Food Preservation Are Molds, Yeasts And Bacteria

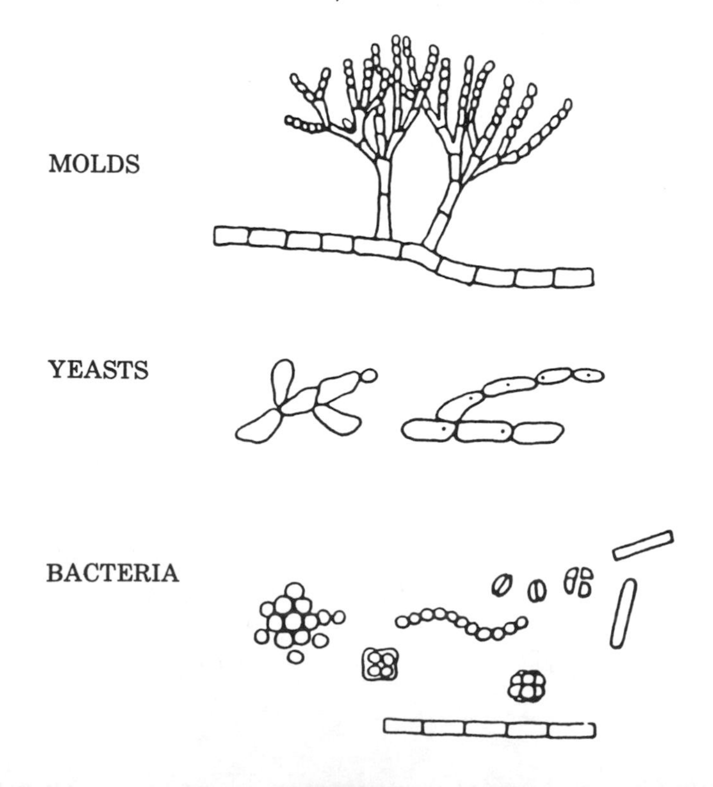

For each bacterial group there is an optimum temperature range for growth. Thermophilic organisms grow at high temperatures. Mesophillic organisms grow at moderate temperatures, while psyshrophillic organisms grow best at cold temperatures Table 14.1.

Microorganisms may be classified on the basis of their peculiar growth characteristics. Osmophilic organisms, which cause spoilage in honey, soft centered candies, jams, molasses, concentrated fruit juices, and other highly concentrated sugar solutions are adapted to growth in substrates of high osmotic pressures. Halophilic microorganisms require a minimum concentration of salt for active growth. Generally Halophilic organisms are found principally in marine environments.

A third way to classify microorganisms is based on the type of nutrients they metabolize. Proteolytic microorganisms break down proteins. Lipolytic organisms cause spoilage of lipid containing foods. Saccharolytic microorganisms, or acid forming microorganisms break down pectin. Amylytic microorganisms break down starch to sugars. Cellulytic microorganisms break down cellulose into simpler carbohydrate compounds.

FIGURE 14.2 - Types Of Mold

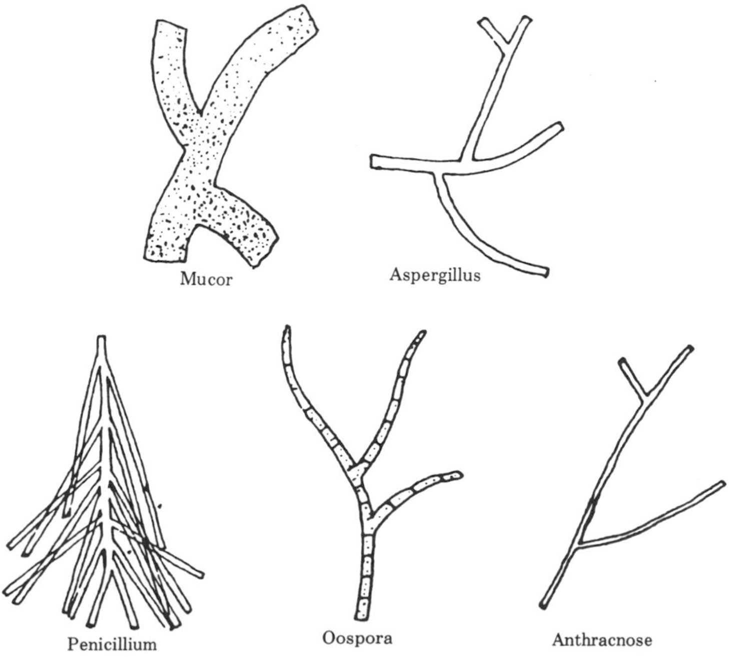

Mucor

Aspergillus

Penicillium

Oospora

Anthracnose

All organisms multiply very rapidly if the conditions for growth are optimum. Studies have shown that a bacterial cell divides about every 20 to 30 minutes. At the end of one hour there are 4 organisms and at the end of 10 hours there may be up to 1,000,000 microorganisms. If the conditions are right in a food plant and there are counts of 75,000 organisms per one square inch, at the end of 3 hours the counts could exceed 4,800,000 organisms (Table 14.2).

Bacteria can be divided into two groups on the basis of their ability to form or not form spores. Most round or cocci and many rod-shaped organisms do not form spores and are called non-spore formers. However, a number of the rod-shaped bacteria have the ability to form spores. A spore is the dormant or resting stage of the microorganism. In the dormant stage the spore can survive a wide range of unfavorable conditions. Generally bacterial spores are resistant to heat, cold, and certain chemical agents. *Clostridium botulinum* is a spore former and this organism produces a deadly toxin if allowed to grow in food. This organism in ubiquitous, that is, it is found everywhere, but the spore will not grow if the pH is below 4.6 or the moisture content (water activity) is below 0.85.

FIGURE 14.3 - Characteristics Of Mold Hyphae

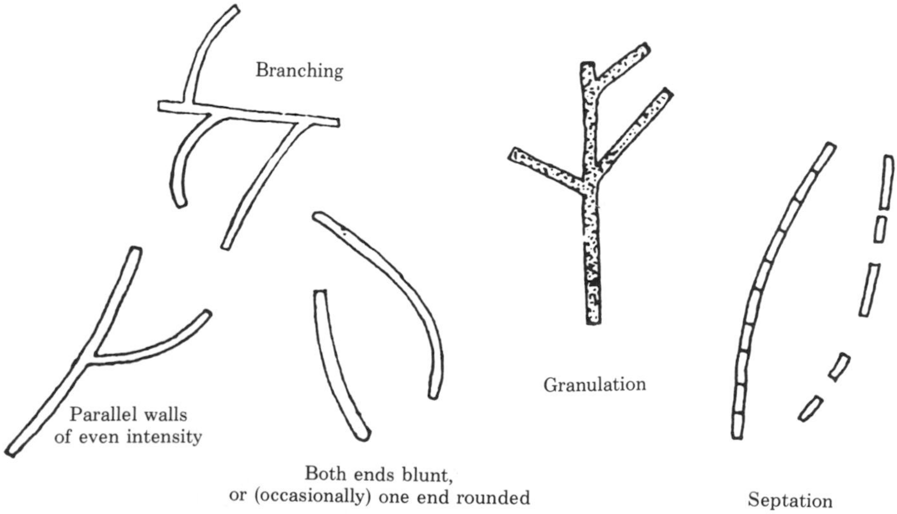

Branching
Granulation
Parallel walls of even intensity
Both ends blunt, or (occasionally) one end rounded
Septation

Six major factors influence the growth of microorganisms: moisture, oxygen concentration, temperature, nutrients, pH, and inhibitors. Each of these factors are complete topics unto themselves and generally are discussed in most microbiology books.

Moisture is a major component of most fresh foods and the amount of water available for microorganisms is referred to as the water activity content (A_w). Water activity is a measure of the free moisture in a product and is the quotient of the water vapor pressure of the substance divided by the vapor pressure of pure water at the same temperature. Pure water has a water activity value of 1.00 and is in equilibrium with a relative humidity of 100 percent. Bacteria require more water than yeasts, which require more water than molds (Table 14.3 and 14.4).

All forms of life require oxygen to carry on metabolic activities. Microorganisms may be classified according to their oxygen requirements. Aerobic organisms need oxygen for growth while anaerobic organisms will not grow if oxygen is present. Most organisms are neither aerobic or anaerobic and are called facultative anaerobes as they can tolerate some degree of the presence or absence of oxygen.

pH is a measure of hydrogen ion concentration and it is a key factor in determining the growth of bacterial spores. A food with a pH of 4.6 or less is termed a high acid food. Acid foods will not permit the growth of spores. If the food is above 4.6 the food is called a low-acid food and will not inhibit the growth of bacterial spores. Foods are classified on the basis of pH (Table 14.4). Some foods are acidified to control pH. FDA in CFR 114.80 list the following 5 procedures for attaining an acceptable equilibrium pH level in the final food product as follows:

1. Blanching of the food ingredients in acidified aqueous solutions.

2. Immersion of the blanched food in acid solutions. Although immersion of food in an acid solution is a satisfactory method for acidification, care must be taken to ensure that the acid concentration is properly maintained.

3. Direct batch acidification, which can be achieved by adding a known amount of acid solution to a specified amount of food during acidification.

4. Direct addition of a predetermined amount of acid to individual containers during production. Liquid acids are generally more effective than solid or pelleted acids. Care must be taken to ensure that the proper amount of acid is added to each container.

5. Addition of acid foods to low-acid foods in controlled proportions to conform to specific formulations.

Inhibitors are chemical compounds that can inhibit the growth of microorganisms by preventing their metabolism, by denaturing the protein portion of the cell, or by causing physical damage to the parts of the cell. Sodium benzoate is a good inhibitor.

Recently, the approved "interim" glossary defines "Indicator Organism" as: "A microorganism, group of microorganisms, or component/product of microorganisms whose presence in a food indicates that there is a significant rise: (1) that a pathogen or its toxin of concern may be present; (2) that faulty practices occurred that may adversely affect the safety of the product; or (3) that the food or ingredient may be unsuited for its intended use." It defines "Microbiological Criteria" as: "Quantitative tolerance on various types, groups, and/or products or microorganisms that cannot be present or exceeded in a food or class of related foods. Typically, these limits are subclassified as standards, guidelines, or specifications depending on the specific regulatory requirements and end-use of the criteria".

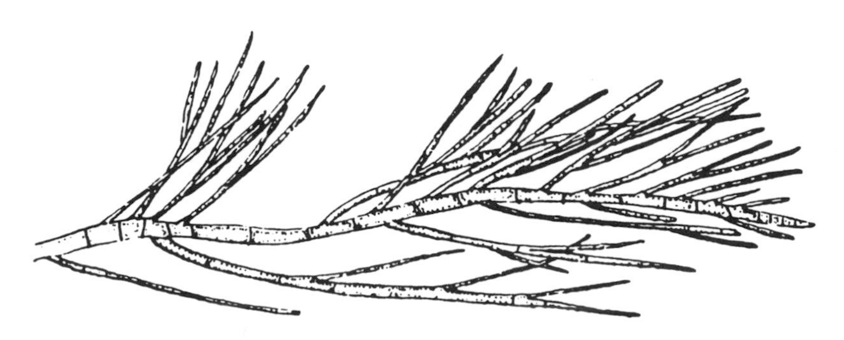

FIGURE 14.4 - Machinery Mold

FIGURE 14.5 - *Geotrichum SP.*

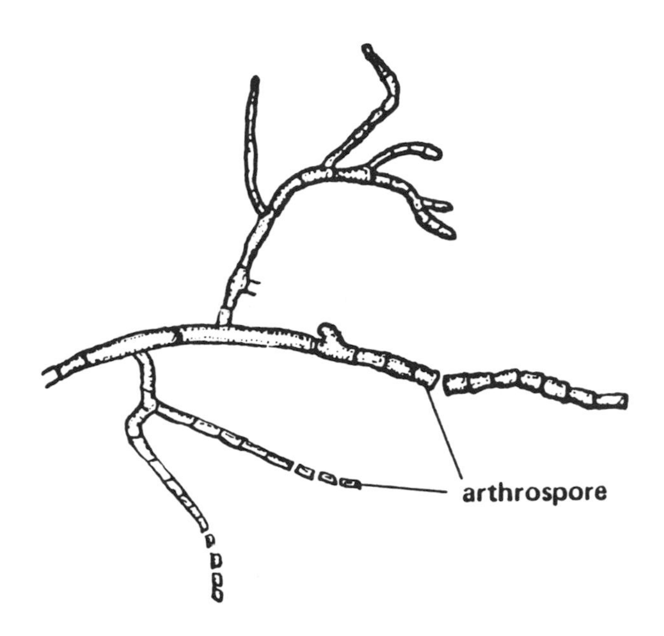

Corlett in a presentation of the FPI's Establishing Hazard Analysis Critical Control Point Programs in 1988 classified micro-organisms on the basis of hazards as follows:

Severe Hazards:

Clostridum botulinum
Shigella
Listeria monocytogenes
Salmonella typhi: paratyphi A & B
Brucella abortus: suis
Mycobacterium bovis
hepatitis A virus
fish and shellfish toxins
certain mycotoxins (aflatoxin)

Moderate Hazards with Potentially Extensive Spread:

Salmonella
pathogenic *Escherichia coli (PEC)*
Streptococcus pyogenes

Moderate Hazards with Limited Spread:
- *Staphylococcus aureus*
- *Clostridium perfringens*
- *Bacillus cerus*
- *Virio parahaemolyticus*
- *Coxiella burnetti*
- *Yersinia enterocolitica*
- *Campylobacter fetus*
- *Trichinella spiralis*
- histamine (from microbial decomposition of scombroid fish)

The following are discussions of 5 of the above organisms.

BOTULISM

Organism: *Clostridium botulinum* (produces botulinum toxin)
Source: Soil, meat, poultry, fish, uncooked vegetables and grains, dust is very widespread.
Disease: Gastrointestinal symptoms, weakness, lassitude, dizziness, blurred vision, difficulty in speech and swallowing, paralysis of respiratory muscles and diaphragm, respiratory failure. Symptoms usually develop 12 to 36 hours after ingestion of food containing botulinum toxin.
Transmission: Spores must be assumed to be present in frozen or refrigerated foods.

Characteristics of Microorganism:
- produces extremely heat resistant spores and grows anaerobically.
- No growth below pH 4.6 demonstrated in commercial-type food systems.
- Heat destroys toxin - boiling for five minutes.
- Spores may germinate and grow in low-acid frozen or refrigerated foods if other microorganisms are not present (cooked products) under certain abuse conditions. Frozen meat pies held in a warm oven for 24 hours were implicated in botulism poisoning.
- Heat treatment activities spore germination and outgrowth; vegetative cells grow from 10°-12°C to 48°-50°C.

CLOSTRIDIUM PERFRINGENS

Organism: *Clostridium perfringens*
Source: Intestinal tract of man and animals, soil.
Disease: *"Perfringens"* food poisoning; severe diarrhea and lower abdominal cramp pain. Symptoms appear in 8-24 hours; duration of illness is usually 12-24 hours.
Transmission: Ubiquitous in nature. Usually present in raw meat and poultry; contamination from humans and animals (including pets) usually from feces.

Characteristics of Microorganism:
- Is a sporeforming anaerobic bacterium.
- Spores survive up to 100°C; some are usually present in "cooked" food.
- Slow cooling and non-refrigerated storage of cooked meat and poultry products with gravy permit rapid growth to high numbers needed for infection. Grow well in higher portion of sensitive temperature range.
- Growth range 6.5° to 50°C; *optimum growth at 43° to 45°C.*

STAPHYLOCOCCUS AUREUS

Organism: *Staphylococcus aureus* strains that produce enterotoxin.
Source: Man and animals. Present in nasal passages of 50 or more of healthy persons. Also associated with pimples, boils, sore throats, colds, and "post-nasal drip."
Disease: *"Staphylococcal"* food poisoning from heat-stable enterotoxins. Symptoms are nausea, vomiting, retching, abdominal cramping and diarrhea. Onset of symptoms is rapid, usually 1-6 hours. Recovery rapid, usually 1-3 days. Seldom fatal.
Transmission: Food handlers, contaminated equipment.

Characteristics of Microorganism:
- Non-sporeforming cocci killed by mild heat.
- Growth range 6.7° to 45.4°C, *optimum 37° to 40°C.*
- pH growth range 4.5 to 9.3.
- Grow at A_W down to 0.88; enterotoxin as low as 0.90.

- **Very heat stable** enterotoxins; may not be completely inactivated by 121°C.
- Organisms resistant to high salt (up to 15%).
- Requires 500,000 cells per gram food to produce sufficient enterotoxin for illness.

ESCHERICHIA COLI - Enteropathogenic Types

Organism: *Escherichia coli* serotypes that cause invasive-type illness.
Source: Intestinal tract of humans and animals (originarily organism is harmless)
Disease: Certain strains of *E. coli* produce enteric diseases in man. Mild to severe diarrhea with profound dehydration and shock, as well as more serious effects. Symptoms appear in 8-44 hours (toxigenic types) and 8-24 hours (invasive types). Can cause death.
Transmission: Raw meat and poultry, raw milk, unprocessed cheese (?), human contamination - handling and poor hygiene. Cross-contamination to cooked foods.

Characteristics of Microorganism:
- Non-sporeforming bacterium; killed by mild heat, does not survive freezing over long periods of time.
- Grows well in most foods at sensitive temperature ranges.
- Presence is unpredictable and testing is very difficult compared to "usual" *E. coli.*
- Difficult to differentiate from non-pathogenic *E. coli.*

SALMONELLA

Organism: *Salmonella* species
1500 serotypes
Example: *S. typhimurium*
Source: Intestinal tract of man and animals.
Disease: *"Salmonellosis,"* mild to severe gastro-enteritis; may cause death. Symptoms appear in 8 to 72 hours, most often between 20 and 48 hours.

Transmission: Egg products, meat and poultry, milk and milk products, animal by-products, dried coconut, animal feed, contamination from water, humans, rodents, cross-contamination.

Characteristics of Microorganism:
- Non-sporeforming rods – killed by mild heat.
- Growth between pH 4.5-9.0; opt. pH 6.5-7.5
- Grow under either aerobic or anaerobic conditions
- Grow on simple nutrients in sensitive temperature range
- Survive well in frozen or dry state in foods
- Temperature range 5.2° to 47°C; optimum is 35° to 37°C.

LISTERIA MONOCYTOGENES

Source: Cattle, raw milk, and dairy cattle may shed this microorganism in milk, other animals, e.g. swine, poultry.
Disease: Monocytosis. Meningitis in infants. Septicemia, abortion, abscesses, endocarditis, conjunctivitis, flulike illness.
Transmission: Dairy products, eggs. Between 10 and 30% of egg-product plant and slaughterhouse workers harbor *L. monocytogenes.*

Characteristics of Microorganism:
- Small motile rod – a non-sporeformer
- May grow down to 2.8°C and lower; grows well at 4.4°C and 36.7°C.
- HTST pasteurization at present minimum legal time-temperature (71.7°C/161°F for 15 seconds) may not kill all Listeria if encapsulated organisms are present at levels of 10^3 to 10^4 organisms/ml, or higher in milk (FRI, 1986; FDA, 1987).
- May not grow below pH 4.8.
- Withstands repeated freezing and thawing
- Survives in dry state, some cases for years
- Survives 15 months in contaminated silage

The Control of microorganisms starts with the thorough evaluation and control of all incoming materials according to rigid specifications. Only those materials and ingredients that are free of gross contamination are acceptable and permitted in the plant. The evaluation should include all receiving containers, as well as shipping containers, for their compliance.

Secondly, since many foods come directly from the soil or are exposed to the atmosphere, foods need to be washed. Washing will remove many organisms and, if followed by chlorination or sanitizer, the product will be much easier to process. Washing is a simple phenomenon where the product should first be wetted or soaked, preferably in a detergent solution. The dwell time in the solution will depend in great part on the type of organisms, but should range from a few seconds to 2 or more minutes. During the soaking, it is advisable to agitate the food materials using either direct injection of steam or air to roll the products against each other. Following the soak period, the product should be washed using a fine mist spray (Spraying Systems Nozzle 18 Sq. or equivalent) with pressures up to 150 psi and allowing the product to be exposed for 30 seconds or more. Of course, the washing is more efficient if the product turns while under the spray wash to expose all sides or areas of the food material. Thorough cleansing of the food at this point will aid greatly in lowering the microorganism count of the raw food material.

Thirdly, all environmental factors must be controlled within the food plant to prevent microbial growth or build-up on or in the equipment and the facility. Ideally, to control growth of microorganisms during processing keep the food at temperatures in excess of 78°C. or at temperatures below 0°C.

There are many test methods for use in determining the sanitation of a food plant. The most important one is sensory evaluation, that is looking and feeling for gross "soil" on the surfaces of belts and equipment and smelling for any decomposed products. Slime mold is one of the first noticeable indicators of insanitary conditions. Slime mold is referred to as machinery mold or more precisely *Geotricum candidum* (Figure 14.4). Fourthly, the personnel must be trained and completely understand the necessity for obeying strict personal hygenic rules and basic sanitation practices. The health and the cleanliness of all employees working in a food plant is crucial to good sanitation practices.

REFERENCES

Anon. 1988. Canned Food-Principles of Thermal Process Control, Acidification and Container Closure Evaluation. The Food Processors Institute, Washington, D.C.

Gorham, J. Richard. 1978. Training Manual for Analytical Entomology in the Food Industry. FDA Technical Bulletin No. 2, Food and Drug Administration, Washington, D.C.

Gould, Wilbur A. 1988. Total Quality Assurance for the Food Industries. CTI Publications, Inc., Baltimore, MD.

Katsuyama, A. M. 1980. Principles of Food Processing Sanitation. Food Processors Institute, Washington, D.C.

Mountney, G. J. and Gould, W. A. 1988. Practical Food Microbiology and Technology, third. Ed. Published by Van Nostrand Reinhold Co., New York, NY.

TABLE 14.1 - CLASSIFICATION OF BACTERIA ACCORDING TO THEIR TEMPERATURE REQUIREMENTS

	Temperature Required For Growth (C)			
BACTERIAL TYPES	MINIMUM	OPTIMUM	MAXIMUM	GENERAL SOURCES OF THESE BACTERIA
Psychrophilic	0 to 5	15 to 20	30	Water and frozen foods
Mesophilic	10 to 25	30 to 40	35 to 50	Pathogenic and many nonpathogenic bacteria
Thermophilic	25 to 45	50 to 55	70 to 90	Many spore forming bacteria usually from soil and water

TABLE 14.2 - Relationship Of Bacterial Growth To Time
(30 minute reproduction cycle)

TIME (Hour)	NO OF BACTERIAL CELLS
0	1
1	4
2	16
3	64
4	250
5	1,000
6	4,000
7	16,000
8	64,000
9	250,000
10	1,000,000

TABLE 14.3 - Water Activity Values For Growth
Of Specific Microorganisms

Microorganisms	Water Activity Value
Aspergillus mold	0.75
Yeasts	0.88
Clostridium botulinum	0.93
Salmonella	0.93

TABLE 14.4 - pH And Moisture Content Of Foods

Food	pH	Moisture (%)
Apples (whole)	3.4-3.5	78.7
Apple juice	3.3-3.5	-
Apricots (dried)	3.6-3.4	-
Asparagus (green)	5.0-5.8	-
Beans		
Baked	4.8-5.5	-
Green	4.9-5.5	95.1
Lima	5.4-6.3	80.7
Soy	6.0-6.6	-
With pork	5.1-5.8	-
Beef (corned, hash)	5.5-6.0	55.6
Beef (stew)	5.6-6.2	72.6
Beef (dried)	5.5-6.8	59.3
Beets (whole)	4.9-5.6	85.4
Blackberries	3.0-4.2	78.4
Blueberries	3.2-3.6	87.9
Boysenberries	3.0-3.3	78.5
Bread		
White	5.0-6.0	-
Date and nut	5.1-5.6	-
Broccoli	5.2-6.0	-
Carrots (chopped)	5.3-5.6	91.0
Carrot juice	5.2-5.8	-
Catfish	-	81.2
Cheese		
Parmesan	5.2-5.3	-
Roquefort	4.7-4.8	-
Cherry juice	3.4-3.6	82.0
Chicken (roasted)	6.2-6.4	47.2
Chicken (with noodles)	6.2-6.7	-
Chop suey	5.4-5.6	-
Cider	2.9-3.3	-
Clams	5.9-7.1	83.7
Cod fish	6.0-6.1	82.6
Corn-on-the-cob	6.5-6.6	76.0
Corn		
Cream style	5.9-6.5	-
Whole kernel		
Brine packed	5.8-6.5	-
Vacuum packed	6.0-6.4	-
Crab apples (spiced)	3.3-3.7	-

TABLE 14.4 - pH And Moisture Content Of Foods - Continued

Food	pH	Moisture (%)
Crackers	7.0-8.5	-
Cranberry		
Juice	2.5-2.7	-
Sauce	2.3-2.3	62.1
Currant juice	3.0-3.0	-
Dates	6.2-6.4	-
Duck (roasted)	6.0-6.1	46.2
Figs	4.9-5.0	87.4
Flour	6.0-6.5	-
Frankfurters	6.2-6.2	-
Fruit cocktail	3.6-4.0	-
Goose (roasted)	-	57.5
Gooseberries	2.8-3.1	-
Grapefruit		
Juice	3.0-3.3	-
Pulp	3.4-3.4	-
Sections	3.0-3.5	-
Grapes	3.5-4.5	74.8
Ham (spiced)	6.0-6.2	-
Herring	-	68.3
Hominy (lye)	6.9-7.9	82.6
Huckleberries	2.8-2.9	90.4
Jams (fruit)	3.5-4.0	-
Jellies (fruit)	3.0-3.5	-
Kidney (beef)	-	76.0
Lamb (cutlets)	-	65.3
Lemons	2.2-2.6	-
Lemon juice	2.2-2.4	-
Liver (calf)	-	66.8
Loganberries	2.7-3.3	-
Mackerel	5.9-6.2	70.1
Milk		
Cow	6.4-6.8	87.0
Evaporated	5.9-6.3	-
Molasses	5.0-5.4	-
Mushrooms	6.0-6.5	92.4
Olives (ripe)	5.9-7.3	78.0
Orange juice	3.0-4.0	-
Oysters	6.3-6.7	-
Peaches	3.4-4.2	88.0
Pears (Bartlett)	3.8-4.6	88.0

TABLE 14.4 - pH And Moisture Content Of Foods - Continued

Food	pH	Moisture (%)
Peas	5.6-6.5	82.9
Pheasant (baked)	-	51.5
Pickles		
Dill	2.6-3.8	-
Sour	3.0-3.5	-
Sweet	2.5-3.0	-
Pimento	4.7-5.2	-
Pineapple		
Crushed	3.2-4.0	80.0
Sliced	3.5-4.1	-
Juice	3.4-3.7	-
Plums	2.8-3.0	-
Pork (leg)	-	53.7
Potatoes		
White	5.4-5.9	72-84
Mashed	5.1-5.1	-
Potato salad	3.9-4.6	-
Prune juice	3.7-4.3	-
Pumpkin	5.2-5.5	84.4
Rabbit	-	75.0
Raspberries	3.2-3.7	-
Rhubarb	2.9-3.3	94.8
Salmon	6.1-6.3	72.0
Sardines	5.7-6.6	53.6
Sauerkraut	3.1-3.7	93.2
Juice	3.3-3.4	-
Shrimp	6.8-7.0	59.4
Soups	-	-
Bean	5.7-5.8	-
Beef broth	6.0-6.2	-
Chicken noodle	5.6-5.8	-
Clam	-	-
chowder	5.6-5.9	-
Duck	5.0-5.7	-
Mushroom	6.3-6.7	-
Noodle	5.6-5.8	-
Oyster	6.5-6.9	-
Pea	5.7-6.2	-
Tomato	4.2-5.2	-
Turtle	5.2-5.3	-
Vegetable	4.7-5.6	-

TABLE 14.4 - pH And Moisture Content Of Foods - Continued

Food	pH	Moisture (%)
Spinach	5.1-5.8	92.1
Squash	5.0-5.3	-
Strawberries	3.1-3.5	-
Sweet potatoes	5.5-5.6	66.7
Tomato	4.1-4.4	42-95
Juice	3.9-4.4	-
Tuna	5.9-6.1	54.1
Turnip greens	5.4-5.6	80.0
Veal	-	75.0
Vegetable		
Juice	3.9-4.3	-
Mixed	5.4-5.6	-
Vinegar	2.4-3.4	-
Youngberries	3.0-3.7	-

CHAPTER 15

WATER AND FOOD PLANT SANITATION

Without question water is one of the most important single ingredients in most food processing operations. Water is a major component of most foods. Water is the vehicle for transfer of power and heat, for conveying materials throughout the plant, and water is used in many food processing operations. Water is, also, used in many unit operations in the preparation of the product for processing. Water may be removed in processing or more water may be added to the product as part of the packing medium for most canned foods. Water is used for cleaning the equipment, utensils, containers, floors, and other processing areas. Water is used by personnel for drinking, washing, and flushing of toilets.

The quantity of water used in food processing varies with the commodities being processed and by the type of processing operation. Murray and Peterson published the data shown in Table 15.1 which is used to illustrate the wide range of water used in the canning industry.

The sources of water present one of the primary problems of water quality. Water may come from lakes, rivers, streams, and canals. This type of water has been classed as Surface Water. Surface water is really run-off water and generally requires treatment by coagulation, sediment removal, followed by filtration and sanitization. This type of water can be difficult and very variable to handle to assure its safety.

Water may be pumped from wells at varying depths and is classed as Ground Water. Ground water can also be potentially dangerous due to seepage from cesspools, sewers, land fills, and other potential sources of contamination.

Sea Water, where applicable, is used in certain parts of the country. If taken near the shore line, it can be of very poor quality due to pollution. Some communities now use reverse osmosis to remove some constituents and impurities from sea water.

TABLE 15.1 - Canning Water Demand
(Taken from Murray and Peterson)

Product	Can Size	Gallons of Water per Case
Asparagus	No. 2, 2 Tall, 300	65-190
Green & Wax Beans	No. 2	45-55
Beets	No. 2	40-50
Carrots	No. 2	40-55
Corn, Cream Style	No. 2	40-50
Corn, Cream Style and Whole Kernel	12 oz., 303, No.2	25-82
Whole Kernel Corn with Peppers	12 oz.	50-82
Lima Beans	No.2	40-55
Mixed Vegetables	No.2	50-60
Peas	303, No. 2, No. 2T	31-135
Pumpkin	No. 2½	60-165
Spinach and Greens	No.2, 2½	75-260
Tomatoes	No. 2½	50-66
Apples	No. 10	75-150
Apricots	No. 2½	50-150
Cherries, Sweet	No. 2½	90-180
Peaches	No. 2½	30-320
Pears	No. 2½	25-180
Plums	No. 2½	50-150
Misc. Fruits	No. 2½	50-100
Misc. Fruits & Vegetables		60
Baked Beans	28 oz.	85
	14 oz.	43
Hominy	No. 10, 2½, 2	55-70
Brown Bread		58
Beans, Soaked Dry	No.2, 2½	30-123

The most common and generally the safest source of water is from municipalities. Generally, the municipality has softened, purified, and chlorinated the water. It is the most reliable source, but may be the most costly.

Depending on the water source, it may carry dissolved minerals, undissolved solids, organic matter in suspension, or dissolved gases such as oxygen and carbon dioxide and various types of microorganisms. The minerals of primary concern to the food processor are calcium, magnesium, sodium, iron, sulfur and manganese. Standards for quality are established by the U.S. Public Health Service and include the items as shown in Table 15.2.

The hardness of water is generally based on the calcium and magnesium salts; however, iron and manganese may show up as part of the hardness factor in water quality. The hardness ions

TABLE 15.2 - Water Quality Tolerances

Characteristic	Maximum Limit in ppm
PHYSICAL	
Turbidity (Silica scale)	10
Color (Platinum scale)	10-20
Objectionable Taste & Odor	0
CHEMICAL	
Arsenic	0.05
Barium	1.0
Cadmium	0.010
Chromium	0.05
Copper	3.0
Iron and Manganese	0.3
Lead	0.05
Magnesium	125.0
Manganese	0.1
Mercury	0.002
Nitrate (as Nitrogen)	10.0
Selenium	0.01
Silver	0.05
Zinc	15.0
Chloride	250.0
Fluoride	1.5
Sulfate	250.0
Phenolic compounds (phenol)	0.001
Total Solids	500.0
Normal carbonate alkalinity	120.0

are generally expressed as calcium carbonate as shown in Table 15.3. Hard water causes many problems in food operations. Most importantly is the scale formation in pipe lines and boiler tubes and as deposits on equipment. Water hardness must be controlled when used in food preparation and in the processing of many commodities. Further, it is essential that the quality of water be controlled when cleaning the equipment and the factory.

TABLE 15.3 - WATER HARDNESS RATINGS QUALITY

Quality	ppm of Calcium Carbonate
Soft Water	less than 50
Slightly Hard Water	50 to 100
Hard Water	100 to 200
Very Hard Water	Greater than 200

Probably the most important quality factor of water is the microbial load that may be present in the water. Many pathogenic organisms may be found in the water supply and it is difficult to isolate and identify all. The commonly accepted practice is to use coliform organisms as indicator organisms. Coliform organisms are used as they represent fecal type organisms. The coliform test is required for most water supplies. Several analytical methods may be used, but the maximum microbial level results all have to conform to given standards.

The most practical means of controlling the bacterial load is through the use of sanitizing agents. Chlorine gas or other sources of chlorine are the most common sanitizers. FDA in the CGMP's Part 113 states that water must be chlorinated or otherwise sanitized for cooling canals and for recirculated water supplies. Furthermore, there should be a measurable residual of the sanitizer at the water discharge point of the container cooler.

FIGURE 15.1 - "Decagon Water Activity Instrument
(Courtesy Decagon Devices, Inc.)

This philosophy is generally applied throughout the food plant, that is, a measurable residual of chlorine in the water after it has been used and one knows that the added chlorine has left the equipment, product, etc. in a sanitary condition. Chlorine may be purchased as compressed chlorine gas in cylinders and when injected into the water, it is readily available as chlorine. It can be easily controlled and it is relatively inexpensive when one compares chlorine gas to other forms of chlorine on the basis of actual chlorine. Chlorine dioxide has also been found quite satisfactory, but it must be generated on site and it must be thoroughly rinsed from the product or equipment before the product or equipment can be used. Chlorine can be combined with calcium and sodium hydroxides to give compounds referred to as hypochlorites. Other compounds that have found use as sanitizers are iodine, bromine, and quaternary ammonium compounds (QUATS).

In-plant chlorination was started back in the '40s and has proved most acceptable and of great value to various segments of the food industry to keep belts, equipment, and the food plant clean. Lengthy discussions of the subject are available from the references at the end of this chapter. For purposes of understanding in-plant chlorination, one must understand the following terms:

CHLORINE DOSAGE: This is the amount of chlorine that is added to water. It may be expressed as parts per million (ppm), or pounds of chlorine added per 24 hours, etc.

CHLORINE DEMAND: This is the difference between the chlorine demand and the chlorine dosage. The chlorine demand varies with the impurities in the water, the pH, the temperature, and other factors.

RESIDUAL CHLORINE: This is the amount of chlorine remaining after the demand has been satisfied. The goal in chlorination is to make certain that there is a free residual amount of chlorine. Residual chlorine can be evaluated with the use of various titration kits.

BREAK POINT CHLORINATION: When small amounts of chlorine are added to water under controlled conditions, the first chlorine is used up in satisfying the chlorine demand of the water. The point at which the added chlorine exceeds the chlorine demand of the water is called the "break point". Further addition of chlorine beyond the break point will result in free residual chlorine concentration that increases almost in direct proportion to the rate of chlorine application. One must satisfy the break point level before any residual occurs or any effect of chlorination can be expected.

IN-PLANT CHLORINATION: This is the addition of chlorine beyond the break point level to water on a continuous basis resulting in free residuals of 2-7 ppm. This water is used continually during many food plant operations to keep belts, tanks, conveyors, etc. clean and in sanitary conditions by constant spraying, flooding, or dipping of the belts or equipment in the water. Clean up water may have chlorine concentration up to 50 ppm. The Food Processors Institute states the following advantage of the in-plant chlorination system:

1. Use of chlorine prevents or reduces microbial buildup on equipment surfaces.
2. Chlorination permits longer hours of operation and/or reduces labor costs by shortening the time required for plant clean-up.
3. Bacterial counts on the raw and finished product are reduced if the product is washed in chlorinated water.
4. Use of normal strength chlorine solutions reduces corrosion of metal surfaces by preventing growth of microorganisms which product acid.

Chlorine should not be used indiscriminately in food plant sanitation. The following precautions are listed by FPI:
1. It must be determined that the flavor of the product is not adverscly affected by chlorine.
2. Strict measures must be taken to prevent contamination of the chlorinated water with phenols or related compounds. The combination compound, chlorophenol, is detectable as an off-flavor in extremely low concentrations and is independent of the type of food being canned.
3. Brines and syrups going into the product should not be chlorinated. However, they may cause no problem if the chlorine is present in very low concentrations and the brine or syrup is heated before adding it to the container.
4. Frequent tests should be made of the chlorine concentration of the water.
5. Standard industrial safety measures should be used in the handling of chlorine containers and the systems used for injecting chlorine into food processing waters.

FPI discusses the following factors which affect the germicidal action of chlorine, that is, the concentration of the chlorine, the pH of the chlorinated water, the amount of organic and inorganic matter in the water, and the temperature of the water.

When using water for cleaning, there are three key factors that will effect the cleaning ability of water. These are the water temperature, the water pressure and the particle or droplet size.

Generally the higher the temperature, the more effective removal of the soil or the contaminants. Of course, some cleaning problems may not tolerate extremely high temperatures. One should remember that holding water in the temperature range of 25° C to 90° C will enhance the growth of thermophilic organisms. Thus the water should be kept below 25° C or above 90° C.

Water under pressure with the right particle or droplet size has a major impact on removal of contaminants. The data in Chart 15.1 show the impact in pounds as water pressure is increased for various nozzle types. Also, the volume of water in cleaning is greatly influenced by the orifice diameter of the nozzle and the pressure of the water as shown in Table 15.4. Nozzle types may vary by manufactures. Generally they are classed as Knife Type or Flat Spray. Fig. 15.1, Flooding Type Fig. 15.2 and Fullcone type Fig. 15.3. The latter type should be used in cooling operations.

CHART 15.1 - Relationship Between Water Pressure By Nozzle Type (Standard Square Fulljet Nozzle) And impact

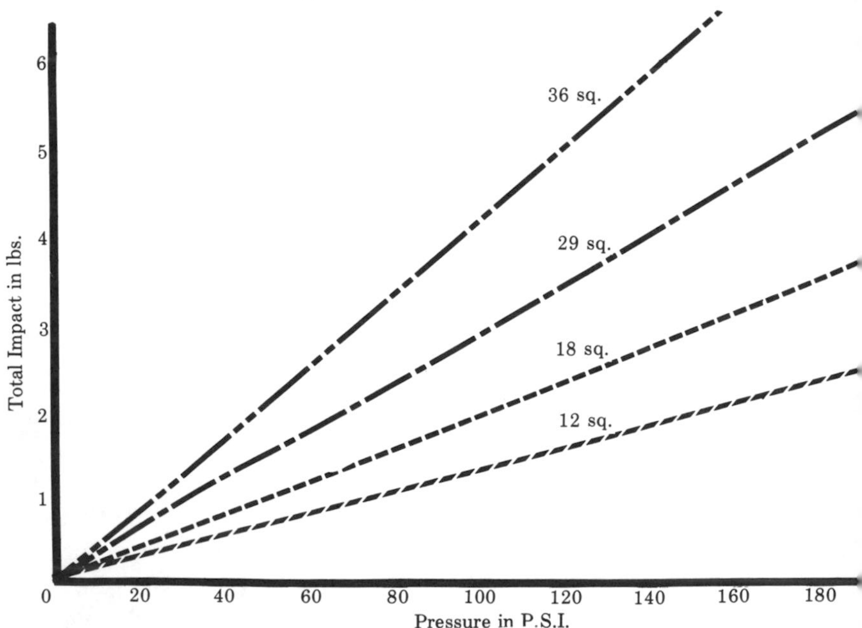

TABLE 15.4 - Relationship Between Full Jet Nozzle By Orifice Opening To GPM At Given Pressure (PSI)*

Pipe Connection	Orifice	Capacity in GPM at given psi						
in inches	(inch)	10	20	40	60	80	100	150
1.4 inch	7/64	1.0	1.4	2.0	2.4	2.8	3.2	3.9
3/8	5/32	1.8	2.5	3.6	4.4	5.1	5.7	7.0
1/2	7/32	2.9	4.1	5.8	7.1	8.2	9.2	11.2
1/2	1/4	3.6	5.1	7.2	8.8	10.2	11.1	14.0

Taken from Spraying Systems Co. Catalogue

FIGURE 15.1 - Flat Spray Nozzle

FIGURE 15.2 - Flooding Type Nozzle

FIGURE 15.3 - Full Cone Type Nozzle

A potential serious problem with water is the possibility of backflow, backsiphonage, and cross connections. Pipe lines should be banded or labeled to indicate what is being carried in all pipe lines. The U.S. Department of Agriculture suggests the following designations:

Fire Lines	Red
Sewer Lines	Black
Edible brine Lines	Green plus name
Inedible brine Lines	Black
Air Lines	White

Line	Color
Potable water Lines	Green
Non potable water Lines	Black
Inedible product Lines	Black plus name
Ammonia Lines	Blue
Edible product Lines	Green plus name
Curing pickle Lines	Green plus name

I would, also, add that steam lines should be labeled, and color coded as Silver.

Pipe lines must be properly installed for safety and prevention of potential contamination. "S" traps, fixed air gap devices, and atmospheric vacuum breakers must be used as required by code. Dead ends are one of the major problem areas for contamination and must be eliminated in all lines.

CHILL A MIRROR TO DEW POINT

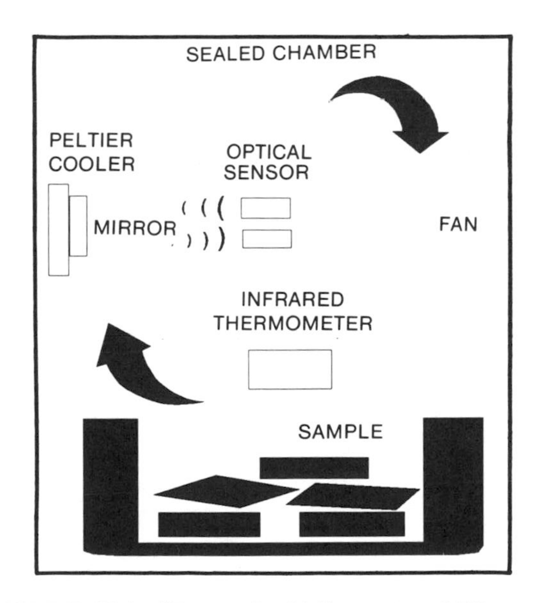

FIGURE 15.4 - "Principle Of Operation Of Decagon Water Activity Instrument"

REFERENCES

Anon, 1988 Canned Foods Principles of Thermal Process Control, Acidification and Container Closure Evaluation. Food Processors Institute, Washington, D.C.

Gould, Wilbur A. 1954. Plant Sanitation Forms Integral Part in Effective Quality Control Program. Food Packer, September.

Gould, Wilbur A. 1956. Pure Water Essential in Cleaning Raw Products. Food Packer, September.

Gould, Wilbur A. 1964. What Every Processor Should Know About Washing Fruits and Vegetables. Canner/Packer, August 25-27.

Katsuyama, Allen M. 1980. Principles of Food Processing Sanitation. Food Processors Institute, Washington, D.C.

Murray, R. V. and G. T. Peterson. 1951. Water for Canning. Continental Can Company, Inc., New York, N.Y.

CHAPTER 16

FUNDAMENTALS OF CHEMICAL CLEANING COMPOUNDS

Dirt or "soil", as the sanitation people call it, has been defined as matter out of place. Thus, cleaning is the removal of matter out of place.

The composition of cleaning compounds, their concentration and the cleaning method depend on the type of soil or dirt on the surface or substance to be cleaned. Food varies widely due to its composition and the changes that may take place in that food during heating and preparation for preservation. Further, foods vary greatly as their solubility.

Most plant food products are made up of carbohydrates and carbohydrates are generally water soluble and easy to remove unless the sugars become caramelized during heating. Animal products, on the other hand, are made primarily of proteins and the proteins are water insoluble, but soluble in alkalies and slightly acid solutions. They are generally difficult to remove and as they denature upon heating they become progressively more difficult to remove. The other major component of many foods is fats or lipids. These substances are likewise water insoluble, but soluble in alkali solutions. The fats will polymerize or breakdown upon heating and they become more difficult to remove as they continue to polymerize.

There are many functions appropriate to specific cleaning compounds. There is no all purpose cleaning compound. Cleaning compounds are specific and their reactions with the soil must be understood if an effective cleaning job is to be done. In most cases all cleaning compounds must be used with water and generally followed by physical methods. First, the functions of chemical cleaners can be defined as follows:

EMULSIFICATION - Emulsification is the mechanical action of breaking up fats and oils into very small particles which are then uniformly mixed with water. In a stable emulsion, the oil particles are held apart and kept suspended uniformly for long periods of time. This action is mechanical just as is homogenization. Emulsifying agents make this mechanical action easy. Often merely gentle agitation will emulsify fats in the presence of a good emulsification agent.

SAPONIFICATION - Saponification is the chemical reaction between an alkali and an animal or vegetable fat resulting in a soap. This is the same basic process which has been known for centuries in which wood ashes were leached to produce alkali, the alkali combined with fats to produce soap. The advantage of this action in cleaning is due to the solubility of fats by the formation of soap. This soluble soil is easy to remove.

SEQUESTERING/CHELATING - Sequesteration is the removal or inactivation of water hardness constituents by the formation of a soluble complex or chelate. This function produces softened water without objectionable precipitation. This type of sequesteration is usually performed by the polyphosphates and may also be termed inorganic sequestering. Chelation activity performs the same function of sequestering water hardness with the exception that the chelating agent forms a typical ring structure with the water hardness constituents. This type of sequestration may also be termed organic sequestration. Chelating substances are generally used in lower concentration than sequestering substances.

WETTING - Wetting is the action of water in contacting all surfaces of soil or equipment. This action is aided by surface tension reduction. Wetting agents lower the surface tension of water by breaking the lines of force, thereby greatly increasing the ability of water to contact all surfaces.

PENETRATION - Penetration is the action of a liquid entering into porous materials through cracks, pinholes, or small channels. This action could be considered a part of wetting. In order to have good wetting action it is necessary to have good penetration.

DISSOLVING - Dissolving is a chemical reaction which produces water soluble products from water-soluble soil. This reaction is very important since some soils such as alkali deposits (lime) cling tenaciously to surfaces, making their physical removal difficult. The application of acids to such soil provides a chemical means of removing these deposits with ease by converting them to a soluble form.

DISPERSION OR DEFLOCCULATION - This is the action of breaking up aggregates or flocs into individual particles. These many small particles then are more easily suspended and flushed off the equipment.

SUSPENSION - Suspension is the action which holds insoluble particles in solution. This action prevents the settling of solids which might form deposits. It also makes it easy to flush the insoluble particles from the surfaces.

PEPTIZING - Peptizing is the physical formation of colloidal solutions from soils which may be only partially soluble. This action is in part similar to dispersion, but is particularly applicable to materials such as protein soils.

RINSING - Rinsing is the condition of solution or suspension which enables it to be flushed from a surface easily and completely. This action is accomplished by reducing the surface tension of the water used. Insoluble materials must be well suspended and present in minimum quantities to promote good rinsing properties.

One other factor related to chemical cleaners is the removal or inactivation of the hardness of the water, if it is hard. Orthophosphates such as trisodium phosphate and alkalies soften water by precipitating the hardness. The complex or polyphosphates soften water by sequestering the hardness. The chelating agents soften water by forming soluble compounds termed "chelates".

The action of sequestration or chelation is an important function of cleaners because the water is softened without objectionable precipitation.

Cleaning compounds can be broadly classified into five main classes:

ALKALIES - These are soil displacing, emulsifying, saponifying and peptizing agents. Sodium hydroxide or caustic soda (lye) is the most common chemical and is the best saponifying agent. Other alkalies are sodium carbonate (soda ash), sodium metasilicate, and sodium bicarbonate.

PHOSPHATES - These are emulsifying and peptizing soil displacement agents. They are, also, dispersing agents, water softening agents and they prevent soil deposits on the cleaned surfaces. They are good water conditioners and a good source of alkalinity. The most important chemical is trisodium phosphate (TSP). Others are Tetrasodium pyrophosphates, sodium tripolyphosphate, sodium hexametaphosphate and sodium tetraphosphate.

WETTING AGENTS/SURFACTANTS - These chemicals wet or penetrate the soil. They are, also, good emulsifiers, dispersing and suspension agents and generally rinse well from the surface. There are three classes of wetting agents, that is, anionic, non-ionic, and cationic. The sulfated alcohols and alkyl aryl sulfonates are the most common anionic agents while the quaternary ammonium compounds are the best known cationic wetting agents. The non-ionic wetting agents are complex organic chemicals which are relatively new on the market. Many of these substances are liquid, they do not ionize, and they are very compatible with other cleaning compounds. Some wetting agents are good defoamers.

ACID CLEANERS - These chemical cleaners are most useful for cleaning alkaline soils, removing mineral deposits, and acting as water conditioners. There are two types: inorganic and organic. The inorganic acids are generally strong, corrosive and hard to handle and usually find little use in the food plant. The most common inorganic acid is phosphoric acid. The organic acids on the other hand are most useful. They include acetic acid, hydrozyacetic acid, lactic acid, tartaric acid, and gluconic acid. Hydrochloric acid may be used for extremely heavy scale removal.

CHELATING AGENTS - These are relatively new chemicals for use to control mineral deposits, soil displacement by peptizing, and for water softening. They are stable to heat and prevent precipitation of water hardness. The most common compound is ethylene diaminetetra acetic acid.

Chemicals for cleaning are generally formulated for specific cleaning problems and they are based on the following factors:

1. The composition and amount of soil.
2. The nature of the surface to be cleaned, that is, stainless steel, aluminum, plastic, painted surfaces, etc.
3. The method of cleaning available, that is, high pressure, circulating, cleaned in place (CIP), cleaned out of place (COP), foam, spray, and/or by hand.
4. The physical nature of the cleaning compound, that is, solid versus liquid.
5. The quality and quantity of the water available for the cleaning job, and
6. The time and temperature available.

Temperature is extremely significant in a cleaning operation. Increasing the temperature has the following beneficial effects:

a-decreasing the strength of the bond between the soil and the surface,
b-decreasing of the viscosity and increasing turbulent action,
c-increases the solubility of the soluble materials,
d-increases the chemical reaction rates.

However, if the temperature is increased too high, greater cleaning problems may be enacted, that is, more denaturization of the protein with greater adhesion to the surface. All other factors being constant, the longer the time up to a given point the more efficient the cleaning action.

REFERENCE

Gilligan, James A. 1979. The Chemistry of Detergents and Sanitizers. Dubois Chemicals, Cincinnati, Ohio

Gould, Wilbur A. 1955. Detergents and Their Role in Cleaning. Food Packer, June.

Gould, Wilbur A. 1980. Good Manufacturing Practices. Snack Food Association, Arlington, Va.

TABLE 16.1 - DEGREE OF ACTIVITY BY FUNCTIONAL PROPERTIES FOR SPECIFIC CLASSES OF CHEMICAL CLEANING COMPOUNDS

FUNCTIONS	DEGREE OF ACTIVITY[1] BY CLASSES OF CHEMICAL CLEANERS					
	Strong Alkalies	Mild Alkalies	Poly Phosphate	Mild Acids	Strong Acids	Surfactants
Chelating	0	1	4	0	0	0
Saponifying	4	3	0	0	0	1
Wetting	1	2	1	1	0	4
Peptizing	4	3	1	2	3	0
Emulsifying	1	2	2	0	0	4
Dispersing	2	3	1	1	0	3
Dissolving	4	3	2	3	4	1
Rinsing	3	3	2	1	0	4
Corrosion	4	2-3	0	2	4	0

[1]Degree of activity - 4 = Extreme, 3 = High, 2 = Medium, 1 = low and 0 none.

FIGURE 16.1 - TEMPERATURE & CHEMICAL ACTIVITY

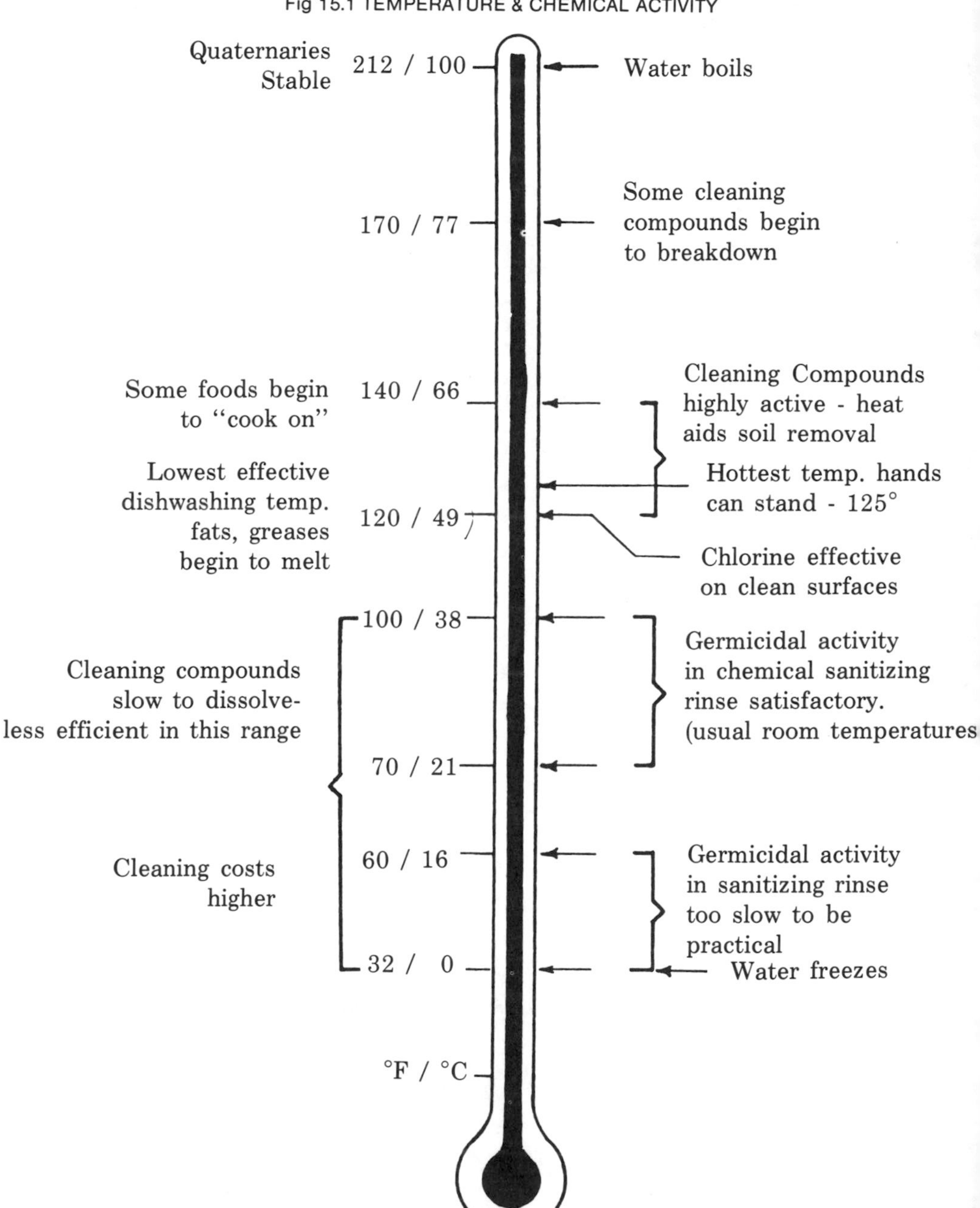

CHAPTER 17

CLEANING THE FOOD PLANT

A clean food plant must be the first goal in producing and processing safe and wholesome foods. This goal must be desired by the firm's management and they must invest the necessary time and money to accomplish it. Secondly, a firm must have properly trained and responsible personnel to maintain the plant and equipment in a clean condition at all times. Thirdly, the sanitation personnel must have the proper tools and materials to accomplish the task of keeping the plant clean. Lastly, the cleaning personnel must know the methods of cleaning each piece of equipment in the food plant and the plant itself. These four elements make up the basic structure of the sanitation program and they must all be put together correctly for an effective sanitation program. Management, men, materials, and methods are the keys for keeping a plant clean and for the production of safe and wholesome foods.

How much to clean, when to clean, where to clean, and what to clean are basic questions that constantly come forth when cleaning a food plant or the equipment therein. These questions should not arise, if the plant has established an organized sanitation program. An organized sanitation program starts with written procedures specifying cleaning chemicals and methods. The written procedures should be developed for every piece of equipment, the plant proper, and all the processes and all the premises.

The cleaning of a food plant or the various unit operations may be continuous throughout the run or day or shift or the cleaning may be accomplished at breaks, shut downs or at the end of the shift. Some food firms use separate sanitation personnel for either practice while others food firms require the production worker at his or her station to maintain their operation in a clean and satisfactory condition. My recommendation is a combination of the above with the thorough cleaning at the end of the day's production. Under this system, the worker should develop pride in maintaining their work station or equipment in a clean condition, but knowing that at the end of the run or shift or day's production, the equipment will be dismantled or cleaned in place for a thorough cleaning by a separate sanitation crew. This system may be the only way to maintain a clean plant and equipment throughout the day's run without serious buildup at any one or more points in the operation.

The established written procedures should detail out the methods of cleaning for each given piece of equipment. That is, will it be cleaned manually (wiped, dusted, brushed, vacuumed, scrubbed, squeeged, etc.) sprayed (high or low pressure), soaked including gel or foam, vacuumed, Cleaned in Place (CIP), or Cleaned out of Place (COP).

Once the method is decided, the next step is to determine the cleaning sequences. Generally in cleaning one should pre-rinse away any debris and gross contamination followed by application of chemical aides (detergents) at given concentration and temperature; followed by rinsing away loosened contamination and soil with pressure (to create a turbulence), nozzle type, and temperature of the water and solution specified; followed by application of appropriate sanitizer; and then followed by post rinsing to remove sanitizer before equipment is reused. These sequences will vary with method of cleaning and type of equipment, but these five steps are the logical steps to follow in cleaning of most equipment, floors, etc.

Burton describes the three criteria that should be used to indicate cleanliness as follows:

A. - Physical cleanliness meaning the absence of visible waste, foreign matter and slime.

B. - Chemically clean meaning removing any undesirable chemical residues. Contamination could come from the cleaning compounds or the sanitizers left on the equipment owing to insufficient rinsing with clean water before reuse.

C. - Microbiologically clean implying freedom from spoilage bacteria as well as organisms that reflect on the general sanitary history for the commodity or process.

The mechanics of soil removal must start with water for many cleaning situations. Water in and of itself will loosen many types of soil and will serve as a vehicle for carrying the soil away. Water may be under low or high pressure depending on the source or changes brought about in the plants system. Water is generally used for clean-up in food plants through the use hoses. All the water hoses should be kept off the equipment and when not in use coiled up and properly hung-up. Short hoses are far superior to long hoses for most food plant operations. All the hoses should have automatic shut-off nozzles. The nozzle may vary in type depending on the piece of equipment to be cleaned. Jet types are used on hard to get at places while flat or full cone nozzles are applicable for cleaning other pieces of

FIGURE 17.1 – Hercules Clean-up Brush
Courtesy Sparta Brush Co., Inc.

equipment. The use of steam injectors into the water lines has advantages for cleaning some pieces of equipment where elevated temperatures will help remove the soil. High pressure water units (stationary or mobile) have found good use for cleaning food plant equipment. Some high pressure units are set-up to pump detergent solutions directly into the lines for additional help in loosening the soil.

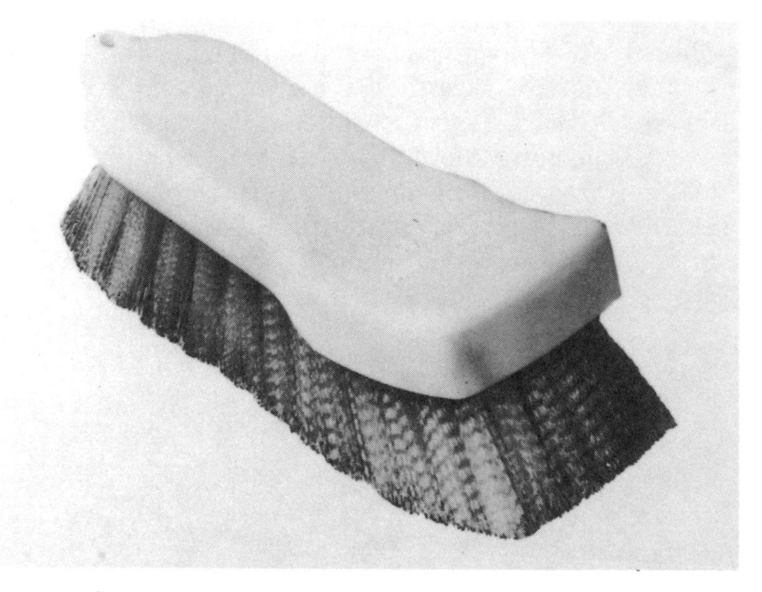

FIGURE 17.2 - Viking General Purpose Brush
Courtesy Sparta Brush Co., Inc.

The mechanics of soil removal may be greatly improved by using appropriate brushes. The brush must be designed for the specific job and it should be impervious to water and the detergent solution. Generally when manually cleaning with brushes, low temperature solutions are required. The brush should be made of nylon or plastic fibers. Wire brushes, straw brushes, plant sponges, or wire sponges shall not be used in food equipment or in food plants. Plastic and stainless steel metal scrappers are acceptable and are useful aides in removing soil. Rubber and plastic squeegees are, also, useful for removing water from equipment and floors. Some equipment, obviously may be cleaned only by wiping. When wiping, clean fiber, paper or cloth towels must be used. They should be single service and an adequate supply should be made available to the sanitation personnel.

FIGURE 17.3 - "Hi-Lo" Floor Brush
Courtesy Sparta Brush Co., Inc.

The mechanics of soil removal is greatly aided by the use of pressure and appropriate spray guns. Low pressure cleaning is quite acceptable for the outside of most food plant equipment with high pressure with or without detergents and/or steam is most useful for the inside of many pieces of food plant equipment. Generally, the detergent solution is made to given concentrations in mixing tanks and then drawn through and mixed with steam by a venturi arrangement and pumped hot under pressure directly on the surface to be cleaned. This arrangement is quite effective for removal of some difficult soils. High pressure cleaning generally has the advantage in removal of soil because only low volumes of water are required.

FIGURE 17.4 – Paddle/Scraper For Cleaning
Courtesy Sparta Brush Co., Inc.

High pressure systems may use only 6 gallons of water per minute and create pressures upwards of 850 psi. High pressure cleaning generally requires less energy, takes less time, and uses less detergent. Further high pressure cleaning permits better soil penetration with more complete capability. As previously indicated, the high pressure system may be portable or a central system can be installed.

Cleaning in Place (CIP) is cleaning the equipment as is through the use of circulating detergent solution through the equipment or by the use of spray balls inside the equipment. Obviously, the equipment cannot be used while it is being cleaned in place. The pumps in the CIP system must be of suf-

FIGURE 17.5 – Cleaning With Foam.
Employees Need Detailed Instructions.

ficient size to cause turbulence of the detergent solution to penetrate the soil. The CIP system allows for thorough mixing of steam or hot water along with the detergent to obtain desired temperature for cleaning the specific equipment. CIP cleaning is a good practice where applicable and requires less labor with repairs kept to a minimum since the equipment is not taken apart. The CIP system requires a series of storage tanks for cleaning and sanitizing solutions, pumps and continuous pipe lines terminating with spray balls or nozzles. The following guidelines should be helpful in designing a CIP system:

1. All surfaces and areas should be accessible to the cleaning solutions.
2. All surfaces in contact with the food or with the detergent solution should be of stainless steel with a finish of at least 2-B or metal of equal resistance and finish.
3. All internal parts should be round or of tubular materials to avoid accumulation of debris and to permit cleaning and flushing away of soil. All welds must be polished and smooth.
4. All pockets, tanks, pipes should be designed to facilitate self-cleaning and self drainage.
5. There shall be no dead ends in any line or piece of equipment.
6. There shall be no interior ledges, recesses, pits, unfinished welds, lips or areas for food or debris to lodge.
7. All joints should be continuously welded and polished smoothly.
8. The CIP pump should be of sufficient size to move the cleaning solutions through the lines and equipment at least 4 times faster than under production runs with food. The pump size will depend on the length of pipe to the equipment or lines to be cleaned and the number of nozzles or balls on any given line.
9. Three to four tanks are required for the cleaning solutions and the sanitizer solutions. Tanks should be of sufficient size to hold up to 50% more solution than needed for cleaning the equipment.
10. Nozzles and spray balls must be designed for each piece of equipment or line to be cleaned.
11. Automatic timers, computers and controllers and diversion valves are essential for efficient operation of and the pumping of the various cleaning solutions through the rinsing cycle.

Cleaning out of Place (COP) is through the use of a washer designed for recirculating cleaning solutions in a tank under high turbulence for cleaning small parts like elbows, filler pieces, valves, pipes, etc. that are difficult to clean. All the pieces must be disassembled to insure proper cleaning of the internal surfaces.

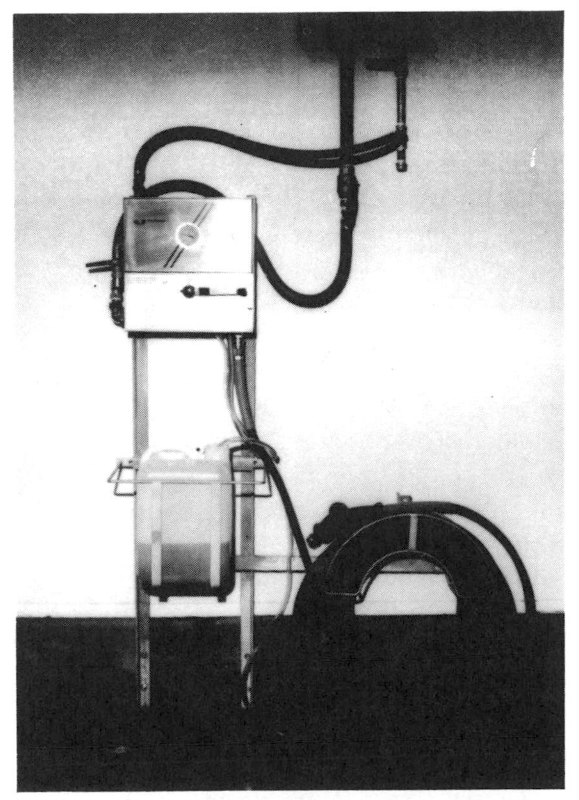

FIGURE 17.6 - High Pressure Hose System

Foam and Gel cleaning is an alternative to soaking and then may be followed by manually brushing or rinsing. It is the application of liquid foam or gel with the detergent chemical entrapped in air to a given surface. The foam or gel is applied to the surface with a pump and pressurized air with appropriate wand directing the foam or gel to the surface to be cleaned. The cleaning action is accomplished by the foam or gel penetrating the soil that is on the surface. The effectiveness of the this type of cleaning is based on the concentration of the cleaning compound, the consistency of the foam or gel, the thickness of the application to the surface. After the foam or gel has broken down the soil, it may be rinsed away leaving the surface clean.

Vacuum cleaning in a food plant is not used as extensively as it should be. This is an efficient method for removing dirt or soil from many pieces of equipment, walls, ceilings, floors, etc. Wet or dry vacuumizing may be effective depending upon the circumstances. Vacuum nozzles should be designed for the given pieces of equipment and hard to get at places. This system of cleaning is far superior to air hoses that just blow the soil from one area to another. Central vacuum lines should be installed with short runs of vacuum hoses to given pieces of equipment. A central collector outside the plant makes for an efficient system of cleaning.

After food plant equipment is cleaned, it should be sanitized. The sanitizer should be strong enough to kill all vegetative microorganisms. The sanitizer should be rinseable from the equipment or surface before it is used. The sanitizer should be non-corrosive to the equipment or surface and it should be safe for the employee to use. The sanitizer should be soluble in water and compatible with other cleaning compounds. Further, the sanitizer should have no detrimental effect on the food being processed. CFR lists in Section 1178.1010 Title 21 the formulation for sanitizers and the maximum permitted levels (see below). Of course, gaseous chlorine can be used as well as edible organic acids and strong alkalis as sanitizers as previously discussed.

With any sanitizer and with all chemicals used in cleaning, the sanitarian must constantly test for the concentration of the chemical in question. All detergent supply houses should provide the customer with testing procedures for any chemical compound purchased from them and they should give complete instructions on the use of their chemicals for cleaning and sanitizing. Further, every chemical distributor must provide the buyer with Material Safety Data Sheets (MSDS). These sheets (1) identify the chemical, who makes it, their address, emergency phone number and date the sheet is prepared; (2) The Hazardous

Ingredients including the formulation, chemical name and ID, and worker exposure limits; (3) Physical and Chemical characteristics of the chemical; (4) Physical Hazards such as fire and explosion; (5) Reactivity data, that is, is the material stable and how to handle it; (6) Health Hazards plus Emergency and First Aid procedures; (7) Precautions for Safe Handling and Use; and (8) Control Measures to reduce harmful exposure. The MSDS gives the user everything they need to work safely with these chemicals. MSDS is only one of the requirements that the user needs to know when working with chemicals. Every food firm must make these sheets available to all employees and the food firm must make sure that users of these chemicals understand the MSDS sheets and how to put the information to use.

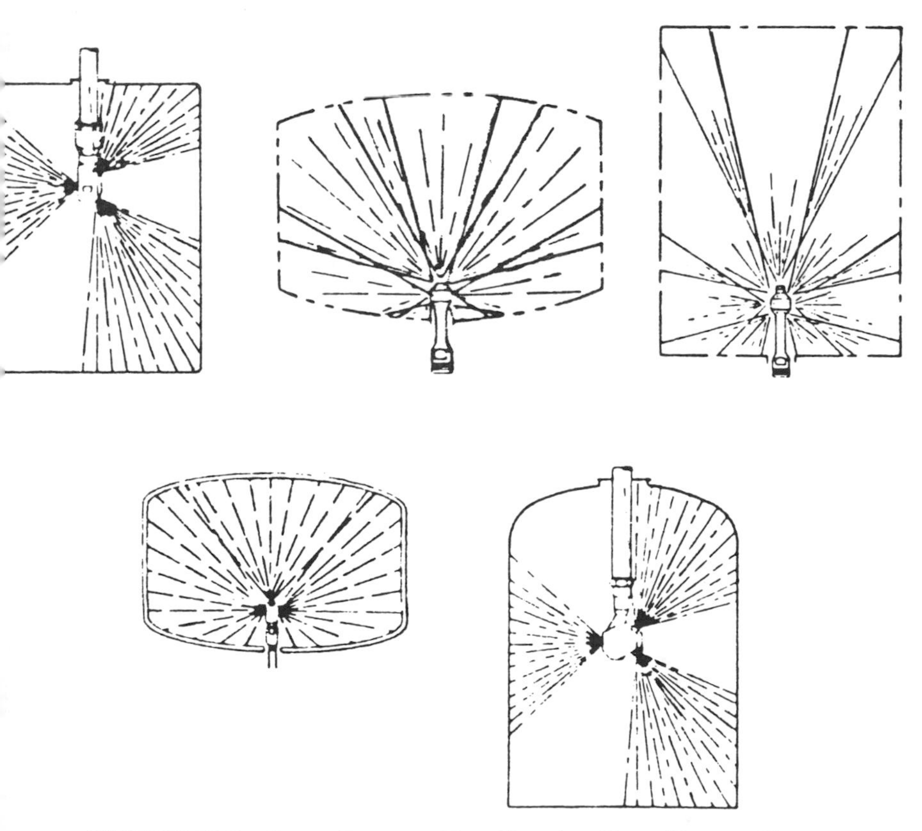

FIGURE 17.7 - Spray Patterns For Cleaning Fans, Drums, Etc.

§ 178.1010 Sanitizing solutions.

Sanitizing solutions may be safely used on food-processing equipment and utensils, and on other food-contact articles as specified in this section, within the following prescribed conditions:

(a) Such sanitizing solutions are used, followed by adequate draining, before contact with food.

(b) The solutions consist of one of the following, to which may be added components generally recognized as safe and components which are permitted by prior sanction or approval.

(1) An aqueous solution containing potassium, sodium, or calcium hypochlorite, with or without the bromides of potassium, sodium, or calcium.

(2) An aqueous solution containing dischloroisocyanuric acid, trichlor-oisocyanuric acid, or the sodium or potassium salts of these acids, with or without the bromides of potassium, sodium, or calcium.

(3) An aqueous solution containing potassium iodide, sodium p-toluenesulfonchloroamide, and sodium lauryl sulfate.

(4) An aqueous solution containing iodine, butoxy monoether of mixed (ethylene-propylene) polyalkylene glycol having a cloudpoint of 90°-100°C in 0.5 percent aqueous solution and an average molecular weight of 3,300, and ethylene glycol monobutyl ether. Additionally, the aqueous solution may contain diethylene glycol monoethyl ether as an optional ingredient.

(5) An aqueous solution containing elemental iodine, hydriodic acid, a-(p-nonylphenyl)-**omega**-hydroxypoly-(oxyethylene)(complyingwiththe identity prescribed in §178.3400(c) and having a maximum average molecular weight of748) and/or polyoxyethylene-polyoxypropylene block polymers (having a minimum average molecular weight of 1,900). Additionally, the aqueous solution may contain isopropyl alcohol as an optional ingredient.

(6) An aqueous solution containing elemental iodine, sodium iodide, sodium dioctylsulfosuccinate, and polyoxyethylene-polyoxypropylene block polymers (having a minimum average molecular weight of 1,900).

(7) An aqueous solution containing dodecylbenzenesulfonic acid and either isopropyl alcohol or polyoxyethylene-polyoxypropylene block polymers (having a minimum average molecular weight of 2,800). In addition to use on food-processing equipment and utensils, this solution may be used on glass bottles and other glass containers intended for holding milk.

(8) An aqueous solution containing elemental iodine, butoxy monoether of mixed (ethylene-propylene) polyalkylene glycol having a minimum average molecular weight of 2,400 and **a**-lauroyl-**omega**-hydroxypoly (oxyethylene) with an average 8-9 moles of ethylene oxide

and an average molecular weight of 400. In addition to use on food processing equipment and utensils, this solution may be used on beverage containers, including milk containers or equipment. Rinse water treated with this solution can be recirculated as a preliminary rinse. It is not to be used as final rinse.

(9) An aqueous solution containing n-alkyl ($C_{12}C_{16}$) benzyldimethylammonium chloride compounds having average molecular weights of 351 to 380. The alkyl groups consist principally of groups with 12 to 16 carbon atoms and contain not more than 1 percent each of groups with 8 and 10 carbon atoms. Additionally, the aqueous solution may contain either ethyl alcohol or isopropyl alcohol as an optional ingredient.

(10) An aqueous solution containing trichloromelamine and either sodium lauryl sulfate or dodecyl-benzenesulfonic acid. In addition to use on food processing equipment and utensils and other food-contact articles, this solution may be used on beverage containers except milk containers or equipment.

(11) An aqueous solution containing equal amounts of n-alkyl ($C_{12}C_{16}$) benzyl dimethyl ammonium chloride and n-alkyl ($C_{12}C_{16}$) dimethyl ethylbenzyl ammonium chloride (having an average molecular weight of 384). In addition to use on food-processing equipment and utensils, this solution may be used on food-contact surfaces in public eating places.

(12) An aqueous solution containing the sodium salt of sulfonated oleic acid, polyoxyethylene-polyoxypropylene block polymers (having an average molecular weight of 2,000 and 27 to 31 moles of polyoxypropylene). In addition to use on food-processing equipment and utensils, this solution may be used on glass bottles and other glass containers intended for holding milk. All equipment, utensils, glass bottles, and other glass containers treated with this sanitizing solution shall have a drainage period of 15 minutes prior to use in contact with food.

(13) An aqueous solution containing elemental iodine and alkyl ($C_{12}C_{16}$) monoether of mixed (ethylene-propylene) polyalkylene glycol,having a cloud-point of 70°-77°C in 1 percent aqueous solution and an average molecular weight of 807.

(14) An aqueous solution containing iodine, butoxy monoether of mixed (ethylene-propylene) polyalkylene glycol, having a cloud-point of 90°-100°C in 0.5 percent aqueous solution and an average molecular weight of 3,300, and polyoxyethylene-polyoxypropylene block polymers (having a minimum average molecular weight of 2,000).

(15) An aqueous solution containing lithium hypochlorite.

(16) An aqueous solution containing equal amounts of *n*-alkyl ($C_{12}C_{16}$) benzyl ammonium chloride (having average molecular weights of 377 to 384), with the optional adjuvant substances tetrasodium ethylenediamin-etetraacetate and/or *alpha*-(*p*-nonylphenol)-*omega*-hydroxy poly(oxyethylene) having an average poly- (oxyethylene) con-

tent of 11 moles. Alpha-hydro-omega-hydroxypoly-(oxyethylene) poly(oxypropoylene) (15 to 18 mole minimum) poly(oxyethylene) block copolymer, having a minimum molecular weight of 1,900 (CAS Registry No. 9003-11-6) may be used in lieu of alpha- (*p*-nonylphenol)-omega-hydroxy- poly(oxyethylene) having an average poly(oxyethylene) content of 11 moles. In addition to use on food-processing equipment and utensils, this solution may be used on food-contact surfaces in public eating places.

(17) An aqueous solution containing di-*n*-alkyl(C_8C_{10})dimenthyl ammonium chlorides having average molecular weights of 332-361 and either ethyl alcohol or isopropyl alcohol. In addition to use on food-processing equipment and utensils, this solution may be used on food-contact surfaces in public eating places.

(18) An aqueous solution containing *n*-alkyl($C_{12}C_{16}$) benzyldimethylammonium chloride, sodium metaborate, *alpha*-terpineol and *alpha*[p-l,l,3,3-tetramethylbutyl)phenyl/]-*omega*-hydroxy-poly (oxyethylene) produced with one mole of the phenol and 4 to 14 moles ethylene oxide.

(19) An aqueous solution containing sodium dichloroisocyanurate and tetrasodium ethylene-diaminetetraacetate. In addition to use on food processing equipment and utensils, this solution may be used on food-contact surfaces in public eating places.

(20) An aqueous solution containing *ortho*-phenylphenol, *ortho*-benzyl-para-chlorophenol, para-tertiaryamylphenol sodium-*alpha*-alkyl($C_{12}C_{15}$)-*omega*-hydroxypoly (oxyethylene) sulfate with the poly(oxyethylene) content averaging one mole, potassium salts of coconut oil fatty acids, and isopropyl alcohol or hexylene glycol.

(21) An aqueous solution containing sodium dodecylbenzenesulfonate. In addition to use on food processing equipment and utensils, this solution may be used on glass bottles and other glass containers intended for holding milk.

(22) an aqueous solutin containing (1) di-*n*-alkyl(C_8C_{10}) dimethylammonium chloride compounds having average molecular weights of 332-361, (2) *n*-alkyl ($C_{12}C_{16}$) benzyldimethylammonium chloride compounds having average molecular weights of 351-380 and consisting principally of alkyl groups with 12 to 16 carbon atoms with or without not over 1 percent each of groups with 8 and 10 carbon atoms, and (3) ethyl alcohol. The ratio of compound (1) to compound (2) is 60 to 40.

(23) An aqueous solution containing *n*-alkyl ($C_{12}C_{16}$) benzyl-dimethyl-ammonium chloride and didecyldimethylammonium chloride.

(24) An aqueous solution containing elemental iodine (CAS Reg. No. 7553-56-2), *alpha*-[p-(1,1,3,3-tetramethylbutyl)-phenyl]-*omega*-hydroxypoly-(oxyethylene) produced with one mole of the phenol and 4 to 14 moles ethylene oxide, and *alpha*-alkyl($C_{12}C_{16}$)-*omega*-hydroxy[poly(oxyethylene) poly(oxypropylene)] (having an average molecular weight of 965).

(25) An aqueous solution containing elemental iodine (CAS Reg. No. 7553-56-2), potassium iodide (CAS Reg. No. 7681-11-0), and isopropanol (CAS Reg. No. 67-63-0). In addition to use on food processing equipment and utensils, this solution may be used on beverage containers, including milk containers and equipment and on food contact surfaces in public eating places.

(26) [Reserved]

(27) An aqueous solution containing decanoic acid (CAS reg. No. 334-48-5), octanoic acid (CAS Reg. No. 124-07-2), and sodium 1-octanesulfonate (CAS Reg. No. 5324-84-5). Additionally, the aqueous solution may contain isopropyl alcohol (CAS Reg. no. 67-63-0) as an optional ingredient.

(28) An aqueous solution containing sulfonated 9-octadecenoic acid (CAS Reg. No. 68988-76-1) and sodium xylenesulfonate (CAS Reg. No. 1300-72-7).

(29) An aqueous solution containing dodecyldiphenyloxidedisulfonic acid (CAS Reg. No. 30260-73-2), sulfonated tall oil fatty acid (CAS Reg. No. 68309-27-3), and neo-decanoic acid (CAS Reg. No. 26896-20-8). In additoin to use on food processing equipment and utensils, this solution may be used on glass bottles and other glass containers intended for holding milk.

(30) An aqueous solution containing hydrogen peroxide (CAS Reg. No. 7722-84-1), peracetic acid (CAS Reg. No. 79-21-0), acetic acid(CAS Reg. No. 64-19-7, and 1-hydroxyethylidene-1,1-diphosphonic acid (CAS Reg. No. 2809-21-4).

(31) An aqueous solution containing elemental iodine, *alpha*-alkyl($C_{10}C_{14}$)-*omega*-hydroxypoly-(oxyethylene) poly(oxypropylene) of average molecular weight between 768 and 837, and *alpha*-alkyl($C_{12}C_{16}$)-omega-hydroxypoly(oxyethylene) poly(oxypropylene) of average molecular weight between 950 and 1,120. In addition to use on food processing equipment and utensils, this solution may be used on food contact surfaces in public eating places.

(32) An aqueous solution containing (i) di-*n*-alkyl(C_8C_{10}) dimethylammonium chloride compounds having average molecular weights of 332 to 361, (ii) n-alkyl($C_{12}C_{16}$)benzyldimethylammonium chloride compounds having average molecular weights of 351 to 380 and consisting principally of alkyl groups with 12 to 16 carbon atoms with no more than 1 percent of groups with 8 and 10, (iii) ethyl alcohol, and

(iv) *alpha*-(p-nonylphenyl)-*omega*-hydroxypoly(oxyethylene) produced by the condensation of 1 mole of *p*-nonylphenol with 9 to 12 moles of ethylene oxide. The ratio of compound (i) to compound (ii) is 3 to 2.

(33) An aqueous solution containing (i) di-*n*-alkyl-(C_8C_{10})-dimethylammonium chloride compounds having molecular weights of 332 to 361; (ii) *n*-alkyl($C_{12}C_{16}$)-benzyldimethylammonium chloride compounds having molecular weights of 351 to 380 and consisting principally of alkyl groups with 12 to 16 carbon atoms with no more than 1 percent of the groups with 8 to 10; and (iii) tetrasodium ethylenediamine tetraacetate. Additionally, the aqueous solution containseither *alpha*-(p-nonyl-phenyl)-*omega*-hydroxypoly-(oxyethylene) or *alpha*-alkyl($C_{11}C_{15}$)-*omega*-hydroxypoly-(oxyethylene), each produced with 9 to 13 moles of ethylene oxide. The ratio of compound (i) to compound (ii) is 3 to 2.

(c) The solutions identified in paragraph (b) of this section will not exceed the following concentrations:

(1) Solutions identified in paragraph (b)(1) of this section will provide not more than 200 parts per million of available halogen determined as available chlorine.

(2) Solutions identified in paragraph (b)(2) of this section will provide not more than 100 parts per million of available halogen determined as available chlorine.

(3) Solution identified in paragraph (b)(3) of this section will provide not more than 25 parts per million of titratable iodine. The solutions will contain the components potassium iodide, sodium *p*-toluenesulfonchloramide and sodium lauryl sulfate at a level not in excess of the minimum required to produce their intended functional effect.

(4) Solutions identified in paragraph (b)(4), (5), (6), (8), (13), and (14) of this section will contain iodine to provide not more than 25 parts per million of titratable iodine. The adjuvants used with the iodine will not be in excess of the minimum amounts required to accomplish the intended technical effect.

(5) Solutions identified in paragraph (b)(7) of this section will provide not more than 400 parts per million dodecylbenzenesulfonic acid and not more than 80 parts per million of polyoxyethylene-polyoxypropylene block polymers (having a minimum average molecular weight of 2,800) or not more than 40 parts per million of isopropyl alcohol.

(6) Solutions identified in paragraph (b)(9) of this section shall provide when ready to use no more than 200 parts per million of the active quarternary compound.

(7) Solutions identified in paragraph (b)(10) of this section shall provide not more than sufficient trichloromelamine to produce 200 parts per million of available chlorine and either sodium lauryl sulfate at a level not in excess of the minimum required to produce its intended function-

al effect or not more than 400 parts per million of dodecylbenzenesulfonic acid.

(8) Solutions identified in paragraph (b)(11) of this section shall provide, when ready to use, not more than 200 parts per million of active quaternary compound.

(9) The solution identified in paragraph (b)(12) of this section shall provide not more than 200 parts per million of sulfonated oleic acid, sodium salt.

(10) Solutions identified in paragraph (b)(15) of this section will provide not more than 200 parts per million of available chlorine and not more than 30 ppm lithium.

(11) Solutions identified in paragraph (b)(16) of this section shall provide not more than 200 parts per million of active quaternary compound.

(12) Solutions identified in paragraph (b)(17) of this section shall provide, when ready to use, a level of 150 parts per million of the active quaternary compound.

(13) Solutions identified in paragraph (b)(18) of this section shall provide not more than 200 parts per million of active quaternary compound and not more than 66 parts per million of *alpha*[p-(1,1,3,3-tetramethylbutyl) phenyl]-*omega*-hydroxypoly (oxyethylene).

(14) Solutions identified in paragraph (b)(19) of this section shall provide, when ready to use, a level of 100 parts per million of available chlorine.

(15) Solutions identified in paragraph (b)(20) of this section are for single use applications only and shall provide, when ready to use, a level of 800 parts per million of total active phenols consisting of 400 parts per million *ortho*-phenylphenol, 320 parts per million *ortho*-benzyl-*para*-chlorophenol and 80 parts per million *para*-tertiaryamylphenol.

(16) Solution identified in paragraph (b)(21) of this section shall provide not more than 430 parts per million and not less than 25 parts per million of sodium dodecylbenzenesulfonate.

(17) Solutions identified in paragraph (b)(22) of this section shall provide, when ready to use, at least 150 parts per million and not more than 400 parts per million of active quaternary compound.

(18) Solutions identified in paragraph (b)(23) of this section shall provide at least 150 parts per million and not more than 200 parts per million of the active quaternary compound.

(19) Solutions identified in paragraphs(b)(24) and (b)(25) of this section shall provide at least 12.5 parts per million and not more than 25 parts per million of titratable iodine. The adjuvants used with the iodine shall not be in excess of the minimum amounts required to accomplish the intended technical effect.

(20)-(21) [Reserved]

(22) Solutions identified in paragraph (b)(27) of this section shall provide, when ready to use, at least 109 parts per million and not more than 218 parts per million of total active fatty acids and at least 156 parts per million and not more than 312 parts per million of the sodium 1-octanesulfonate.

(23) Solutions identified in paragraph (b)(28) of this section shall provide, when ready to use, at least 156 parts per million and not more than 312 parts per million of sulfonated 9-octadeceonoic acid, at least 31 parts per million and not more than 62 parts per million of sodium xylenesulfonate.

(24) Solutions identified in paragraph (b)(29) of this section will provide at least 237 parts per million and not more than 474 parts per million dodecyldiphenyloxidedisulfonic acid, at least 33 parts per million and not more than 66 parts per million sulfonated tall oil fatty acid, and at least 87 parts per million and not more than 174 parts per million neodecanoic acid.

(25) Solutions identified in paragraph (b)(30) of this section shall provide, when ready to use, not less than 550 parts per million and not more than 1,110 parts per million hydrogen peroxide, not less than 100 parts per million and not more than 200 parts per million peracetic acid, not less than 150 parts per million and not more than 300 parts per million acetic acid, and not less than 15 parts per million and not more than 30 parts per million 1-hydroxyethylidene-1,1-diphosphonic acid.

(26) The solution identified in paragraph (b)(31) of this section shall provide, when ready to use, at least 12.5 parts per million and not more than 25 parts per million of titratable iodine. The adjuvants used with the iodine will not be in excess of the minimum amounts required to accomplish the intended technical effect.

(27) Solutions identified in paragraph (b)(32) of this section shall provide, when ready to use, at least 150 parts per million and no more than 400 parts per million of active quarternary compounds in solutions containing no more than 600 parts per million water hardness. The adjuvants used with the quarternary compounds will not exceed the amounts required to accomplish the intended technical effect.

(28) Solutions identified in paragraph (b)(33) of this section shall provide, when ready to use, at least 150 parts per million and not more than 400 parts per million of active quaternary compounds. The adjuvants used with the quaternary compounds shall not exceed the amounts required to accomplish the intended technical effect. Tetrasodium ethylenediamine tetraacetate shall be added at a minimum level of 60 parts per million. Use of these sanitizing solutions shall be limited to conditions of water hardness not in excess of 300 parts per million.

(d) Sanitizing agents for use in accordance with this section will bear labeling meeting the requirements of the Federal Insecticide, Fungicide, and Rodenticide Act.

[42 FR 14609, Mar. 16, 1977, as amended at 42 FR 37974, July 26, 1977; 42 FR 44544, Sept. 6, 1977; 43 FR 1941, Jan. 13, 1978; 45 FR 25389, Apr. 15, 1980; 45 FR 56797, Aug. 26, 1980; 46 FR 31006, June 12, 1981; 46 FR 60567, Dec. 11, 1981; 47 FR 53848, Nov. 30, 1982; 48 FR 6705, Feb. 15, 1983; 48 FR 23179, May 24, 1983; 48 FR 30613, July 5, 1983; 50 FR 23949, June 7, 1985; 50 FR 30148, July 24, 1985; 50 FR 40965, Oct. 8, 1985; 51 FR 7437 and 7438, Mar. 4, 1986; 51 FR 33892, Sept. 24, 1986; 51 FR 47226, Dec. 31, 1986; 52 FR 409, Jan. 6, 1987]

CHAPTER 18

EMPLOYEES AND FOOD PLANT SANITATION

The food plant employee is a vital part of the operation of any food firm. Management must indoctrinate the employee before they are hired on the fundamental principles of food plant sanitation and personal hygiene. Management must require a pre-employment physical examination to assure itself that the potential employee is in good mental, physical, and emotional health. Further, management should insist that the employee follow good personal health practices and sound hygiene while in its employ. Management must impress on each employee the need for food plant sanitation practices while at his station of work. Finally, management must emphasize the basic food laws and regulations and offer refresher sessions in food plant sanitation on an annual basis to all food handlers.

Employees are the key to productivity and compliance to the CGMP's in any food firm. Their actions, habits, and attitudes can directly affect the outcome of the operation. People must have innate sanitation habits to be successful employees in a food firm. They must want to live and work in a clean environment and they must want to constantly learn how to improve their habits to always produce products that are safe and clean. Only by people caring about cleanliness and safety can the industry survive and fulfill its intended purpose. Sanitation is a way of life and it must come from within the individual. Their attitude is a reflection of their desire and interest in promoting good sanitary practices. People tend to emulate their superiors. Thus, superiors, including all in management must set the example for personal cleanliness, personal habits, and personal attitudes. This leadership will encourage employee cooperation and the proper development of good sanitary practices at all times.

The following is taken from the CGMP's Part 110.10 and clearly sets forth the rules and regulations for all food plant personnel.

"**Personnel** - The plant management shall take all reasonable measures and precautions to ensure the following:

(a) *Disease control.* Any person who, by medical examination or supervisory observation, is shown to have, or appears to have, an illness, open lesion, including boils, sores, or infected wounds, or any other abnormal source of microbial contamination by which there is a reasonable possibility of food, food-contact surfaces, or food packaging materials becoming contaminated, shall be excluded from any operations which may be expected to result in such contamination until the condition is corrected. Personnel shall be instructed to report such health conditions to their supervisors.

(b) *Cleanliness.* All persons working in direct contact with food, food contact surfaces, and food packaging materials shall conform to hygienic practices while on duty to the extent necessary to protect against contamination of food. The methods for maintaining cleanliness include, but are not limited to:

(1) Wearing outer garments suitable to the operation in a manner that protects against the contamination of food, food contact surfaces, or food packaging materials.

(2) Maintaining adequate personal cleanliness.

(3) Washing hands thoroughly (and sanitizing if necessary to protect against contamination with undesirable microorganisms) in an adequate hand washing facility before starting work, after each absence from the work station, and at any other time when the hands may have become soiled or contaminated.

(4) Removing all insecure jewelry and other objects that might fall into food, equipment, or containers and removing hand jewelry that cannot be adequately sanitized during periods in which food is manipulated by hand. If such hand jewelry cannot be removed, it may be covered by materials which can be maintained in an intact, clean, and sanitary condition and which effectively protects against the contamination by these objects of

the food, food-contact surfaces, or food packaging materials.

(5) Maintaining gloves, if they are used in handling, in an intact, clean, and sanitary condition. The gloves should be of an impermeable materials.

(6) Wearing, where appropriate, in an effective manner, hair nets, head bands, caps, beard covers, or other effective hair restraints.

(7) Storing clothing or other personal belongings in areas other than where food is exposed or where equipment or utensils are washed.

(8) Confining the following to areas other than where food may be exposed or where equipment or utensils are washed: eating food, chewing gum, drinking beverages, or using tobacco.

9) Taking any other necessary precautions to protect against contamination of food, food-contact surfaces, or food packaging materials with microorganisms or foreign substances including, but not limited to perspiration, hair, cosmetics, tobacco, chemicals, and medicines applied to the skin.

c) *Education and training.* Personnel responsible for identifying sanitation failures or food contamination should have a background of education or experience, or a combination thereof, to provide a level of competency necessary for production of clean and safe food. Food handlers and supervisors should receive appropriate training in proper food handling techniques and food-protection principles and should be informed of the danger of poor personal hygiene and insanitary practices.

(d) *Supervision.* Responsibility for assuring compliance by all personnel with all requirements of this part shall be clearly assigned to competent supervisory personnel".

The responsibility for compliance rests with the supervisor and he or she must be qualified to interpret the above for management. However, management must see that all personnel receive necessary training in proper food handling techniques and food-protection principles and the danger of poor personal hygiene and insanitary practices.

FIGURE 18.1 – The Food Plant Workers Are Trained On A Regular Basis In The Principles Of Food Plant Sanitation And The CGMP's

Management must provide clean rest rooms, clean dressing and change rooms and/or locker rooms, and hand washing facilities in the rest room and near the work station with liquid soap, sanitizer, and single service towels or air driers. Further, for some operations, management should provide sufficient uniforms with appropriate laundry service.

Management must provide appropriate training for all food handling personnel in food sanitation principles and practices. The training program must be on-going and each employee should be required to attend training sessions on an annual basis. Many up to-date sanitation practices and training programs are available on film, video, slides and tapes. Many firms have their own in-house programs tailored to their operation. These are most effective when using the firm's individual plant worker along with management. The Food and Drug Administration has excellent flyers and posters to help in training programs.

All employees must understand basic personal hygiene rules and regulations. The following are additional personal hygiene points and other practices to be followed and enforced by the firm's management:

1. Daily bathing of all personnel.
2. Washing of hair, at least, once per week.
3. Keeping nails clean and properly trimmed.
4. Keeping underclothing clean and keeping uniforms clean.
5. Wearing hair nets to restrain all hair.
6. Males must be clean shaven or they must wear hair restraints or facial snoods. Further, a mustache must be well trimmed and never below the corners of the mouth.
7. Males should not have side burns below ear lobes and no "mutton chop" side burns.
8. All personnel should learn to wash hands after:
   - -coughing and sneezing,
   - -visiting the toilet,
   - -after smoking,
   - -after breaks,
   - -before returning to work station,
   - -handling soiled containers or waste materials,
   - -handling animal products,

   and, after using the telephone.
9. Pens, pencils, etc. shall not be carried in pockets above the waist line. Preferably no garment shall have pockets above the waist line.
10. Glass bottles, cups, glasses, beakers, and other glass containers shall not be permitted in the food preparation, processing or packaging areas, unless used for packaging of food.
11. Safe personal conduct within the food plant shall be strictly observed. Running, horseplay, riding on trucks or lifts, taking shortcuts (ducking under conveyors, etc. whether operating or not) are prohibited,

12. Protective shoes and clothing, including goggles where appropriate, should be worn at all times,
13. Each production worker must be responsible for maintaining his or her work area from undue accumulations of food, dust, dirt, or waste in which insects or bacteria may harbor or breed,
14. All employees must flush urinals and toilets after each usage,
15. All doors and windows must be provided with special screening to inhibit ingress of insects or rodents and they should never be left ajar,
16. All intermediate containers should be kept covered when containing products,
17. Sweaters or other 'wooly' substances on employee's person should be avoided in the production, processing, or packaging of food. If a sweater is worn, it should be covered with a non-lint uniform,
18. Fingernail polish should never be worn in the preparation, processing or packaging areas,
19. Maintenance personnel shall not place their tools, parts being repaired, etc. on food contact surface areas,
20. Any food items dropped and coming in contact with the floor must be immediately discarded rather than placed back into production.

Management must make routine inspection of all lockers to be assured that the locker is kept clean and in compliance. The locker is the property of the food firm and they would be held responsible if contamination came from a given locker. Therefore, inspection on a regular routine basis is the only answer.

This list should be expanded by the firm's supervisors to cope with the situation in the facility. These suggestions along with the CGMP's in Part 110.10 should be given to every employee and the employee should be asked to read them and sign a statement that he or she has read them and will comply. Many sanitation problems could be solved if all firms followed this practice.

Another suggestion deals with hazards in the food plant. The employee, under the Right to Know Act and the Occupational Safety and Health Administration's regulations is entitled to know the written hazard communication program in existence by the firm. Further they are entitled to know the location of the Material Safety Data Sheets as previously discussed. The following is a Supervisors Hazard Communication Check List as prepared by the Food Processors Institute.

FORM 18.1 - Employee Training And Information

Check ✓ Only if Answer is Yes

- ☐ 1. I know the location of the written Hazard Communication Program and Material Safety Data Sheets.
- ☐ 2. I have read the written program and understand it.
- ☐ 3. I fully understand the purpose of the program and my related responsibilities to the program.
- ☐ 4. I know which hazardous chemicals are used or stored in the work area I supervise.
- ☐ 5. Material Safety Data Sheets (MSDS) are on file for each hazardous chemical in my work area.

EACH MSDS CONTAINS THE FOLLOWING:

- ☐ 6. The identity (chemical or brand name) used on the label.
- ☐ 7. The chemical and common name (unless protected by trade
- ☐ secret provisions of the law) for each chemical.
- ☐ 8. Physical and chemical characteristics of the chemical.
- ☐ 9. The chemical's physical or health hazards.
- ☐ 10. Primary route of entry.
- ☐ 11. Permissible exposure limits, where available.
- ☐ 12. Whether the chemical has been listed as a potential carcinogen.
- ☐ 13. Any known precautions for safe handling or applicable control measures.
- ☐ 14. Emergency and first aid procedures.
- ☐ 15. Date of MSDS preparation at last change.
- ☐ 16. Name, address, and telephone number where further information can be obtained.

LABELS

- ☐ 17. All hazardous chemical containers used or stored in my area of supervision are clearly labeled.
- ☐ 18. Labels include chemical identity, hazard warnings,and name and address of the manufacturer.
- ☐ 19. Portable, unlabeled containers are not used for storage.
- ☐ 20. I am aware of any chemicals that may pass through unlabeled pipes.

EMPLOYEE TRAINING AND INFORMATION

- ☐ 21. The employees I supervise have been informed of the purpose and location of the written hazard communication program and the Material Safety Data Sheet (MSDS).
- ☐ 22. Newly assigned employees or employees in an area where a new hazard has been introduced are informed of the hazard communication requirements and all operations in the work area where hazardous chemicals are present.
- ☐ 23. Hazardous chemicals in the work area have been identified.
- ☐ 24. Hazards of nonroutine tasks and the hazards of chemicals in unlabeled pipes, as well as appropriate precautions to be taken, have been fully explained to employees.
- ☐ 25. Employees can read and understand container labels.
- ☐ 26. I have explained how to read and understand the MSDS.
- ☐ 27. Employees have been trained in methods and observations that are used to detect the presence or release of hazardous chemicals, including the detection of unusual smells and other warning signs.
- ☐ 28. I have communicated the physical and health hazards of chemicals used in my work area.
- ☐ 29. Employees have been trained and medically qualified to use respiratory protection equipment where required.
- ☐ 30. Employees have been trained in the proper use of protective clothing.
- ☐ 31. The use of protective gear is strictly enforced.
- ☐ 32. Training has included first aid measures and emergency procedures.
- ☐ 33. Methods to inform outside contractors of chemical hazards their employees may be exposed to on site have been developed as well as methods of informing employees under my supervision of hazardous chemicals brought on site by outside contractors.
- ☐ 34. I feel confident that employees under my supervision have been adequately trained and informed according to the hazardous communication standard established by this plant and OSHA.

(Note, any item not checked implies that further information is needed or that steps must be taken to make the program comply with OSHA's regulation).

REFERENCES

Anon. So You Work in a Food Plant. U. S. Department of Health and Human Services, Washington, D. C. HHS Publication No (FDA) 72-2032.

Anon. Supervisor's Hazard Communication Check List. The Food Processors Institute, Washington, D. C.

Gould, Wilbur A. 1977. Good Manufacturing Practices for Snack Food Manufacturers. Potato Chip/Snack Food Association, Arlington, Va.

CHAPTER 19

FOOD SAFETY AND HAZARD ANALYSIS

All consumers expect the food they buy and eat to be safe and wholesome. The food industry has a moral and legal obligation to produce safe and wholesome food. The food industry has a good record, but times have changed in recent years with the new technology and preservation practices, new regulations and amendments to the food laws, and new packaging and marketing practices. Food firm management must continue to be the first line of defense in the production and manufacture of safe and wholesome foods. They must make this first and foremost in their many duties and policies.

The first step in a food plant safety program is the establishment of a safety committee with full authority for assuring the manufacture of safe and wholesome food. The committee should be made up of the following persons or their representatives: Purchasing Officer, Marketing Manager, Plant Engineer, Production Manager, Sanitarian, Personnel Manager and the Quality Assurance Manager. The Quality Assurance manager should chair the committee and report directly to top management. The quality assurance manager must have full authority to control all incoming materials, products in process, all finished products in the warehouse, shipping vehicles and the authority to audit the market place. Only by thorough inspection and evaluation can one hope for safe and wholesome foods reaching the ultimate consumer.

In previous chapters, discussions of some of the major hazards have been discussed, that is, bacteria and microorganisms, insects, birds, and many foreign substances. These and many other hazards are the target of the safety committee.

Food safety rests with everyone in the food chain. The food chain is only as strong and good as the weakest link. It starts with the producer and his responsibilities to use only the pesticides, fertilizers, herbicides, defoliants, growth regulators, and feed that is within the legal limits of the laws and regulations. Every food firm must have a thorough knowledge of his producers and must know how, what, where, when, and how much of any chemical or additive is used. The food firm procurements department must operate with rigid specifications with no side deals or favoritisms or under the table deals. The specifications must include acceptable quality levels, deliveries on time, and vendors that are qualified and reliable. The firm must audit all incoming materials into the factory and the firm must live by quality standards that are objective and within specification limits.

The second major safety cause in any food firm is the methods of manufacture. The food firm must have strict operating specifications for every unit operation in the plant. These include time, temperatures, pressures, flow, etc. All of the operating specifications must be in writing and the operator must be held accountable. All the hazards must be noted and evaluation methods established for control of any potential safety problems (See Figure 19.1). Other hazards that may be critical are paint, lubricants, cleaners, sanitizers, pesticides, metal and other foreign objects. All of these must be under the control of management to assure the safety of the food during process and in the marketing channels.

The use of statistical process control practices is mandatory for the production of uniform quality products and the requirement is an asset to the operator. Within the food plant, all hazards in the process must be identified and control limits must be established for every unit operation. SPC is a practice if adhered to will assure management of products being manufactured in full compliance to given standards and company policy. Through the use of SPC, management will be assured of the elimination of defective and unsafe products. The SPC program must be audited by the quality control personnel and the production manager on a regular and routine basis. The specifications including cleaning standards, safe chemical aides and their parameters, sanitizing practices, and post rinsing schedules prior to reuse.

As important as the above is, the operators and plant employees must operate within the best of personal hygienic practices. Someone has stated that the plant personnel are mirror images of management. Generally, plant personnel take great pride in their work and they will show much enthusiasm if given a voice in the decision process. However, they must be given adequate training and they must be constantly up-dated. All plant personnel must be held accountable for their personal habits and for their working within the standards set forth by management. People are the most important ingredient in the plant and they can be most effective in eliminating hazards by working together in teams or circles to control quality, production, and to assure the manufacture of safe and wholesome foods. They can make great contributions to the success of the food firm by eliminating hazards and any potential safety problems. Personnel must be held fully accountable for their actions and their behaviors and they must be rewarded for their positive efforts. No plant can operate without people and all personnel must comply with all rules and regulations as they too have moral obligations to the public, themselves, and the firm they work for.

The third key safety factor in every food firm is the hazards associated with the various commodities and/or the various processes or preservation methods in use. As previously discussed, foods vary widely in composition, but probably the most significant component of most foods and/or formulated foods is the inherent pH. pH is a measure of hydrogen ion concentration or technically, it is the negative logarithm of the hydrogen ion concentration. Foods with a maximum pH of 4.6 or less are called as high acid foods while foods with a pH greater than 4.6 are low acid foods. Berries, sauerkraut, plums, cherries, and other fruits including tomatoes are examples of high acid foods. Most vegetables, meat, and fish are low acid foods. The pH of a food will influence the types of bacteria that will grow in it. This is most significant in that *Clostridium botulinum* will not grow in food if the pH is below 4.8. In low-acid foods, high heat is required to kill the spores of *C. botulinum* or the spores of other food spoilage organisms. Foods may be acidified for preservation to control the pH. These acidified foods must have a pH of 4.6 or less and a water activity greater than 0.85. Examples are artichoke hearts, acidified bean salads, acidified peppers and/or pimentos, marinated vegetables and fresh pack pickles.

Other methods of food preservation, such as, concentration rely on high solids content or low water activity levels for their preservation. Examples are, jams, jellies, concentrates and dried products.

Frozen foods are preserved by low temperatures where microorganisms generally will not grow. The product must be kept frozen as the organisms may grow if the temperature conditions are elevated.

With today's new technology, the trend is to process foods less and rely on preservation additives, refrigeration, controlled atmospheres and/or controlled packaging to extend the shelf life of many food items. All of these practices must be carefully monitored or potential food safety problems could take place.

Many foods today are so-called formulated or engineered foods and, as such, there may be ingredients mixed into the products that are most sensitive to given health hazards or potential safety problems. The product development managers must be cognizant of all sensitive ingredients and the potential problems they possess. Further, the product development managers must understand preservation methods and apply principles that will control the growth of these microorganisms throughout the process and marketing channels. Their responsibilities cannot be understated from the standpoint of the production of safe and wholesome foods. Product development technologists play an important role in the growth of a food firm and they must be well aware of their responsibilities for the good of all concerned.

In all of the above the quality assurance personnel have major responsibilities as they must uphold the policies of the food firm for the full safety of the foods being manufactured and marketed. Their responsibilities include the development of rigid specifications, inspection of all incoming materials for acceptable quality levels, control of the production lines to given specified limits, auditing of all finished products in the warehouse and the warehouse itself as to time the products are in the warehouse, that the warehouse temperatures are within limits, and the practice of First In First Out (FIFO). Further the quality assurance personnel must audit the market channels with a complete follow up on any possible complaints at the consumer levels. One other area of responsibility for the quality assurance personnel is to be certain that the label is clear and complete in terms of use of the product by the consumer including preparation practices and storage prior to use. The responsibilities of the quality assurance department cannot be over stated. They are the last line of defense for the food firm in assuring safe and wholesome foods.

Since the late '60's, efforts have been put forth to isolate all the potential hazards in the food chain and develop critical control point checks. The USDA in 1989 adopted a Hazard Analysis Critical Control Point (HACCP) system as developed by a National Advisory Committee on Microbiological Criteria for Foods. This document (see below) defines HACCP as a systematic approach to be used in food production as a means to assure food safety. They state that each food establishment must develop their own HACCP system and tailor it to their product, process and distribution system.

In summary, hazards in the food industry may be physical, chemical or microbiological. They present new challenges to the industry and they must be eliminated during the manufacture of safe and wholesome foods. Management must establish firm and rigid policies on all incoming materials with specifications and acceptable quality limits, all equipment and operations in the food plant must operate with process control limits, all personnel must be indoctrinated in hygienic practices and be held accountable for operating within the purview of established standards, all finished products must be audited for quality assurance through the marketing channels, and every firm must establish a quality assurance department with full authority to uphold the policies and standards of the food firm for the production of safe and wholesome products. High quality foods that are safe are in great demand and the technology is available to any food firm to live well within the rules and regulations for the production of foods that are most acceptable. The ultimate result is growth for the industry and great satisfaction to the consumer.

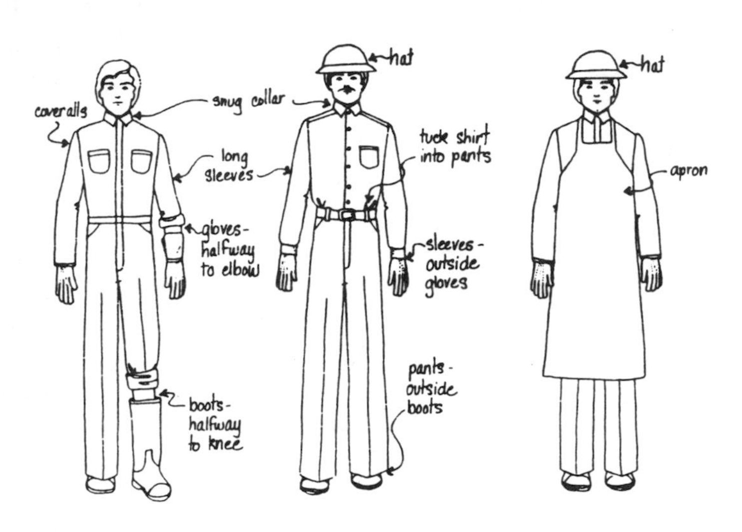

FIGURE 19.1 – Proper Dress For Sanitarian And Most Food Plant Workers

REFERENCES

Anon. 1989 HACCP Principles for Food Production. United States Department of Agriculture, Food Safety and Inspection Service Manual.

Stevenson, Kenneth E. 1993 HACCP Establishing Hazard Analysis Critical Control Point Programs – A Workshop Manual. Food Processors Institute, Washington, DC.

TABLE 19.1 – WAGCO's Potato Salad With Indicated Potential Hazards And Critical Control Points

POTENTIAL HAZARDS	INGREDIENTS				CRITICAL CONTROL POINTS
	Potatoes	Onions	Parsely	Sauce*	
Pesticides, handling, holding, and shipping containers	Recieving[1]	Receiving[1]	Recieving[1]	Recieving[1]	[1]AQL to specifications
Water quality	Washing	Peeling	Washing	Weighing[2]	[2]Formulation
Chemical aides	Peeling	Trimming[3]	Inspecting[3]	Blending	[3]Defect removal
	Trimming[3]	Chopping[4]	Chopping[4]	Heating	[4]Uniformity
	Cutting[4]	Inspecting[3]	Weighing[2]	Pumping	
	Inspecting[2]	Weighing[2]			
Steam Quality	Blanching[5]				[5]Enzyme inactivation
	Weighing[2]				
		Mixing[6]			[6]Time and temperature
Cleaning aides for containers		Filling[7]			[7]Capicity of container and temperature
		Closing[8]			[8]Integrity of seal
		Coding[9]			[9]Line, date, quality
		Processing[10]			[10]Time & temperature
Quality of water and chlorine residual		Cooling[11]			[11]Center temperture and chlorine residual
Time & temperature		Warehousing[12]			[12]FIFO and shelf life

*Sauce is formulated from Stabilized Salad Dressing, Vinegar, Spices, Salt and Water.

HACCP Principles for Food Production

HAZARD ANALYSIS & CRITICAL CONTROL POINT SYSTEM

National Advisory Committee On
Microbiological Criteria For Foods

Adopted November, 1989

EXECUTIVE SUMMARY

In response to a request from the Chairman of the National Advisory Committee on Microbiological Criteria for Foods (Committee), Dr. Lester M. Crawford, an ad hoc working group chaired by Dr. Donald A. Corlett undertook the assignment of drafting a guide setting forth the principles of Hazard Analysis Critical Control Point (HACCP) Systems. The Committee has espoused HACCP as an effective and rational approach to the assurance of food safety. This document represents "HACCP" as used by this Committee. There has been no attempt made to draft a specific HACCP plan for any commodity. HACCP systems must be developed by individual producers and tailored to their individual processing and distribution conditions.

This document defines HACCP as a systematic approach to be used in food production as a means to assure food safety. Seven basic principles underlie the concept. These principles include an assessment of the inherent risks that may be present from harvest through ultimate consumption. Six hazard characteristics and a ranking schematic are used to identify those points throughout the food production and distribution system whereby control must be exercised in order to reduce or eliminate potential risks. A guide for HACCP plan development and critical control point (CCP) identification are noted. Further, the document points out the additional areas that are to be included in the HACCP plant--the need to establish critical limits that must be met at each CCP, appropriate monitoring procedures, corrective action procedures to be taken if a deviation is encountered, recordkeeping, and verification activities.

1.0 Preamble

The National Advisory Committee on Microbiological Criteria for Foods (Committee) endorses the Hazard Analysis and Critical Control Point (HACCP) System as an effective and rational approach to the assurance of food safety. In the application of HACCP, the use of microbiological testing is seldom an effective means of monitoring critical control point (CCP) because of the time required to obtain results. In most instances, monitoring of CCP can best be accomplished through the use of physical and chemical tests, and through visual observations. Microbiological criteria do, however, play a role in verifying that the overall HACCP system is working.

The Committee believes that the HACCP principles should be standardized to create uniformity in its work, and in training and applying the HACCP system by industry and regulatory authorities. In accordance with the National Academy of Sciences recommendation, the HACCP system must be developed by each food establishment and tailored to its individual product, processing and distribution conditions.

2.0 Definitions

2.1 Continuous Monitoring: Uninterrupted recording of data such as a recording of temperature on a strip chart.

2.2 Control Point: Any point in a specific food system where loss of control does not lead to an unacceptable health risk.

2.3 Critical Control Point: Any point or procedure in a specific food system where loss of control may result in an unacceptable health risk.

2.4 Critical Defect: A defect that may result in hazardous or unsafe conditions for individuals using and depending upon the product.

2.5 Critical Limit: One or more prescribed tolerances that must be met to insure that a critical control point effectively controls a microbiological health hazard.

2.6 Deviation: Failure to meet a required critical limit for a critical control point.

2.7 HACCP Plan: The written document which delineates the formal procedures to be followed in accordance with these general principles.

2.8 HACCP System: The result of the implementation of the HACCP principles.

2.9 Hazard: Any biological, chemical, or physical property that may cause an unacceptable consumer health risk.

2.10 Monitoring: A planned sequence of observations or measurements of critical limits designed to produce an accurate record and intended to insure that the critical limit maintains product safety.

2.11 Risk: An estimate of the likely occurrence of a hazard or danger.

2.12 Risk Category: One of six categories prioritizing risk based on food hazards.

2.13 Sensitive Ingredient: Any ingredient historically associated with a known microbiological hazard.

2.14 Significant Risk: Posing moderate likelihood of causing an unacceptable health risk.

2.15 Spot Check: Supplemental tests performed on a random basis.

2.16 Verification: Methods, procedures and tests used to determine if the HACCP system is in compliance with the HACCP plan.

3.0 Purpose and Principles

HACCP is a systemic approach to food safety, consisting of the seven following principles:

3.1 Assess hazards and risks associated with growing, harvesting, raw materials and ingredients, processing, manufacturing, distribution, marketing, preparation and consumption of the food.

3.2 Determine CCP required to control the identified hazards.

3.3 Establish the critical limits that must be met at each identified CCP.

3.4 Establish procedures to monitor CCP.

3.5 Establish corrective action to be taken when there is a deviation identified by monitoring a CCP.

3.6 Establish effective record-keeping systems that document the HACCP plan.

3.7 Establish procedures for verification that the HACCP system is working correctly.

4.0 Explanation of Principles
4.1 **Principle No. 1**: Assess hazards associated with growing, harvesting, raw materials and ingredients, processing manufacturing, distribution, marketing, preparation and consumption of the food.

4.1.1 **Description:** Provides for a systematic evaluation of a specific food and its ingredients or components to determine the risk from hazardous microorganisms or their toxins. Hazard analysis is most useful for guiding the safe design of a food product and defining the CCP that eliminate or control hazardous microorganisms or their toxins at any point during the entire production sequence. The hazard assessment is a two-part process consisting of ranking of food according to six hazard characteristics, followed by the assignment of risk category which is based upon the ranking.

Ranking according to hazard characteristics is based on assessing a food in terms of (a) whether the product contains microbiologically sensitive ingredients, (b) whether the process does not contain a controlled processing step that effectively destroys harmful microorganisms, (c) whether there is significant risk of post processing contamination with harmful microorganisms or their toxins, and (d) whether there is substantial potential for abusive handling in distribution or in consumer handling or preparation that could render the product harmful when consumed or (e) whether there is no terminal heat process after packaging or when cooked in the home.

Ranking according to these six characteristics results in the assignment of risk categories based on how many of the characteristics are present.

The risk categories are utilized for recognizing the hazard risk for ingredients and how they must be treated or processed to reduce the risk for the entire food production and distribution sequence.

The hazard assessment procedure is ideally conducted after developing a working description of the product, establishing the types of raw materials and ingredients required for preparation of the product, and preparing a diagram for the food production sequence. The two-part assessment of hazard analysis and assignment of risk categories is conducted according to the following procedure:

4.1.2 Hazard analysis and assignment of risk categories:

4.1.2.1 Rank the food according to hazard characteristics A through F, using a plus (+) to indicate a potential hazard. The number of pluses will determine the risk category. A model diagram outlining this concept is given under section 4.1.3. As indicated, if the product falls under hazard class A, it should automatically be considered Risk Category VI.

Hazard A:

A special class that applies to non-sterile products designated and intended for consumption by at risk populations, e.g., infants, the aged, the infirm, or immunocompromised individuals.

Hazard B:

The product contains "sensitive ingredients" in terms of microbiological hazards.

Hazard C:

The process does not contain a controlled processing step that effectively destroys harmful microorganisms.

Hazard D:

The product is subject to recontamination after processing before packaging.

Hazard E:

There is substantial potential for abusive handling in distribution or in consumer handling that could render the product harmful when consumed.

Hazard F:

There is no terminal heat process after packaging or when cooked in the home.

Note: Hazards can also be stated for chemical or physical hazards, particularly if a food is subject to them.

4.1.2.2 **Assignment of risk category**
(based on ranking by hazard characteristics):

Category VI.

A special category that applies to non-sterile products designated and intended for consumption by at risk populations, e.g., infants, the aged, the infirm, or immunocompromised individuals. All six hazard characteristics must be considered.

Category V.

Food products subject to all give general hazard characteristics. Hazard class B, C, D, E, F

Category IV.

Food products subject to four general hazard characteristics.

Category III.

Food products subject to three of the general hazard characteristics.

Category II.

Food products subject to two of the general hazard characteristics.

Category I.

Food products subject to one of the general hazard conditions.

Category O.
Hazard Class - No hazard.

Note: Ingredients are treated in the same manner in respect to how they are received at the plant, BEFORE processing. This permits determination of how to reduce risk in the food system.

4.1.3 It is recommended that a chart be utilized that provides assessment of a food by hazard characteristic and risk category. A format for this chart is given as follows:

Food Ingredient or Product (VI,V,IV,III,II,I,O)	Hazard Characteristics (A,B,C,D,E,F)	Risk Category
T	A+(Special Category)*	VI
U	Five +'s (B through F)	V
V	Four +'s (B through F)	IV
W	Three +'s (B through F)	III
X	Two +'s (B through F)	II
Y	One + (B through F)	I
Z	No +'s	O

* Hazard characteristic A automatically is risk category VI, but any combination of B through F may also be present.

4.2 **Principle No. 2:** Determine CCP required to control the identified hazards.

4.2.1 **Description:** A CCP is defined as any point or procedure in a specific food system where loss of control may result in an unacceptable health risk. CCP must be established where control can be exercised. All hazards identified by the hazard analysis must be controlled at some point(s) in the food production sequence, from harvesting and growing raw materials to the ultimate consumption of the food.

CCP are located at any point in a food sequence where hazardous microorganisms need to be destroyed or controlled. For example, a specified heat process, at a given time and temperature to destroy a specified microbiological pathogen, is a CCP. Likewise, refrigeration required to prevent hazardous organisms from growing, or the adjustment of a food to a pH necessary to prevent toxin formation is a CCP.

Types of CCP may include, but are not limited to: cooking, chilling, sanitizing, formulation control, prevention of cross contamination, employee hygiene and environmental hygiene.

CCP must be carefully developed and documented. In addition, they must be used only for purposes of product safety. They should not be confused with control points that do not control safety. For comparison, a control point is defined as any point in a specific food system where loss of control does not lead to an unacceptable health risk.

4.3 **Principle No. 3:** Establish the critical limits which must be met at each identified CCP.

4.3.1 **Description:** A critical limit is defined as one or more prescribed tolerances that must be met to insure that a CCP effectively controls a microbiological health hazard. There may be more than one critical limit for a CCP. If any one of those critical limits is out of control, the CCP will be out of control and a potential hazard can exist. The criteria most frequently utilized for critical limits are temperature, time, humidity, moisture level (A_w), pH, titratable acidity, preservatives, salt concentration, available chlorine, viscosity and in some cases, sensorial information such as texture, aroma and visual appearance. Many different types of limit information may be needed for safe control of a CCP.

For example, the cooking time of meat patties should be designed to eliminate the most heat-resistant vegetative pathogen which could reasonably be expected to be in the product. The critical limits must be specified for temperature, time and meat patty thickness. Technical development of these critical limits requires accurate information on the probably maximum numbers of these microorganisms in the meat, use of additional ingredients and the potential for recontamination.

The relationship between the CCP and its critical limits for the meat patty example is shown as follows:

Critical Control Point	Critical Limits
Meat patty cooked to destroy the most heat resistant pathogen, based on lethality tests. The minimum lethal cook will usually be designated "to reach an internal patty temperature of x for time y."	Minimum operating temperature of cooker to achieve microbiological lethality at center of coldest patty. Time to achieve lethality (belt speed expressed at rpm). Patty thickness. Other possible critical limits: -- oven humidity -- patty composition -- cooker sanitation -- etc.

This example illustrates that the type and number of critical limits will vary depending on the type of cooking system and equipment used for meat patties.

4.4 **Principle No. 4** Establish procedures to monitor CCP.

4.4.1 **Description**: Monitoring is the scheduled testing or observation of a CCP and its limits. Monitoring results must be documented. From the monitoring standpoint, failure to control a CCP is a critical defect.

A critical defect is defined as a defect that may result in hazardous or unsafe conditions for individuals using and depending upon the product. Because of the potentially serious consequences of a critical defect, monitoring procedures must be extremely effective.

Ideally, monitoring should be at the 100% level. Continuous monitoring is possible with many types of physical and chemical methods. For example, the temperature and time for the scheduled thermal process of low-acid canned foods is recorded continuously on temperature recording charts. If the temperature falls below the scheduled temperature or the time is insufficient, as recorded on the chart, the retort load is restrained as a process deviation. Likewise, pH measurement may be done continually in fluids or by testing of a batch before processing. There are many ways to monitor CCP limits on a continuous or batch basis and record the data on charts. The high reliability of continuous monitoring is always preferred when feasible. It requires careful calibration of equipment.

When it is not possible to monitor a critical limit on a full-time basis, it is necessary to establish that the monitoring interval will be reliable enough to indicate that the hazard is under control. Statistically designed data collection systems or sampling systems lend themselves to this purpose. However, statistical procedures are most useful for measuring and reducing the variation in food formulations, manufacturing equipment and measuring devices. Thus, they increase the reliability of the system.

When using statistical process control, it is important to recognize that there is no tolerance for exceeding a critical limit. For example, when a pH of 4.6 or less is required for product safety, no single product unit may have a pH above 4.6. To compensate for variation, the maximum of the product may be targeted at a pH below 4.6. Statistical process control can be applied to understand variation in the system, and assure that no unit exceeds a pH of 4.6. Statistical audits can be based on this concept.

Most monitoring procedures for CCP will need to be done rapidly because they relate to on-line processes and there will not be time for lengthy analytical testing. Microbiological testing is seldom effective for monitoring CCP due to their time-consuming nature. Therefore, physical and chemical measurements are preferred because they may be done rapidly and can indicate microbiological control of the process.

Physical and chemical measurements that may be utilized for monitoring include:

Temperature;

Time;

pH;

Sanitation at CCP;

Specific preventive measures for cross contamination;

Specific food handling procedures;
Moisture level; and
Other.

Spot checks are useful for supplementing the monitoring of certain CCP and their respective limits. They may be used to check incoming pre-certified ingredients, assess equipment and environmental sanitation, airborne contamination, cleaning and sanitizing of gloves and any place where follow-up is needed. Spot check may consist of physical and chemical tests and, where needed, microbiological tests.

With certain foods, microbiologically sensitive ingredients, or imports, there may be no alternative to microbiological testing. However, a sampling frequency that is adequate for reliable detection of low levels of pathogens is seldom possible because of the large number of samples needed. For this reason, microbiological testing has limitations in a HACCP system, but is valuable as a means of establishing and randomly verifying the effectiveness of control at CCP, (challenge tests, spot checking or for troubleshooting.)

All records and documents associated with CCP monitoring must be signed by the person doing the monitoring and signed by a responsible official of the company.

4.5 **Principle No. 5:** Establish corrective action to be taken when there is a deviation identified by monitoring of a CCP.

4.5.1. **Description:** Actions taken must eliminate the actual or potential hazard which was created by deviation from the HACCP plan, and assure safe disposition of the product involved. Because of the variations in CCP for different food and the diversity of possible deviations, specific corrective actions must be developed for each CCP in the HACCP plan. The actions must demonstrate that the CCP has been brought under control. Deviation procedures must be documented in the HACCP plan and agreed to by the appropriate regulatory agency prior to approval of the plan.

Should a deviation occur, the plant will place the product on hold pending completion of appropriate corrective actions and analyses. In instances where is may be difficult to determine the safety of the product, then the testing and final disposition must be agreed to by the government. In instances not associated with safety, government consultation is not required.

Identification of deviant lost and corrective actions taken to assure safety of these lots must be noted in the HACCP record and remain on file for a reasonable period after the expiration date or expected shelf life of the product.

4.6 **Principle No. 6:** Establish effective record-keeping systems that document the HACCP plan.

4.6.1. **Description:** The HACCP plan must be on file at the food establishment. Additionally, it is to include documentation relating to CCP and any action on critical deviations and disposition of product. Those materials are to be made available to government inspectors upon request. The HACCP plan clearly designates records that will be available for government inspection. Certain records that deal with the functioning of the HACCP system and proprietary information are not necessarily available to regulatory agencies.

Generally the types of records utilized in the total HACCP system will include the following:

(Note: Only those records pertaining to CCP must be made available to regulatory agencies.)

4.6.1.1 **Ingredients**

Supplier certification documenting compliance with processor's specifications.

Processor audit records verifying supplier compliance.

Storage temperature record for temperature sensitive ingredients.

Storage time records of limited shelf life ingredients.

4.6.1.2 **Records relating to product safety**

Sufficient data and records to establish the efficacy of barriers in maintaining product safety.

Sufficient data and records establishing the safe shelf life of the product.

Documentation of the adequacy of the processing procedures from a knowledgeable process authority.

4.6.1.3 **Processing**

Records from all monitored CCP.

System records verifying the continued adequacy of the processes.

4.6.1.4 **Packaging**

Records indicating compliance with specifications of packaging materials.

Records indicating compliance with sealing specifications.

4.6.1.5 **Storage and Distribution**

Temperature records.

Records showing no product shipped after shelf life date on temperature sensitive products.

4.6.1.6 **Deviation File**

4.6.1.7 **Modification** to the HACCP plan file indicating approved revisions and changes in ingredients, formulations, processing, packaging and distribution control, as needed.

4.7 **Principle No. 7:** Establish procedures for verification that the HACCP system is working correctly.

4.7.1 **Description:** Verification consists of methods, procedures and tests used to determine that the HACCP system is in compliance with the HACCP plan. Both the producer and the regulatory agency have a role in verifying HACCP plan compliance. Verification confirms that all hazards were identified in the HACCP plan when it was developed. Verification measures may include physical, chemical and sensory methods and testing for conformance with microbiological criteria when established.

4.7.1.1 **Examples of verification activities** include but are not limited to:

Establishment of appropriate verification inspection schedules.

Review of the HACCP plan.

Review the CCP records.

Review deviations and dispositions.
Visual inspections of operations to observe if CCP are under control.
Random sample collection and analysis.
Written record of verification inspections which certifies compliance with the HACCP plan or deviations from the plan and the corrective actions taken.

4.7.1.2 **Verification inspections** should be conducted when:
Routinely, or on an unannounced basis to assure selected CCP are under control.
It is determined that intensive coverage of a specific commodity is needed because of new information on food safety issues requiring assurance that the HACCP plan remains effective.
Foods produced have been implicated as a vehicle of foodborne disease.
Requested on a consultative basis or established criteria have not been met.

4.7.1.3 **Elements** which must be included in verification inspection reports:
Existence of an approved HACCP plan and designation of person(s) responsible for administering and updating the HACCP plan.
All records and documents associated with CCP monitoring must be signed by the person monitoring and approved by a responsible official of the firm.
Direct monitoring data of the CCP while in operation.
Certification that monitoring equipment is properly calibrated and in working order.
Deviation procedures.
Any sample analysis for attributes confirming that CCP are under control to include physical, chemical, microbiological or organoleptic methods.

5.0 Guide for HACCP plan implementation for a specific food:

5.1 Describe the food and its intended use.

5.2 Develop a flow diagram for the production of the food.

5.3 Perform a hazard assessment (Principle 1).
  a. Ingredients prior to any processing step.
  b. End product.

5.4 Select CCP (Principle 2).
  a. Enter a flow diagram in numerical order.
  b. List CCP number and description.

5.5 Establish critical limit (Principle 3).

5.6 Establish monitoring requirements (Principle 4).

5.7 Establish corrective action to be taken when there is a deviation identified by monitoring of a CCP (Principle 5).

5.8 Establish effective recordkeeping systems that document the HACCP plan (Principle 6).

5.9 Establish procedures for industrial and governmental verification that the HACCP system is working properly. Verification measure may include physical, chemical and sensory methods, and when needed, establishment of microbiological criteria (Principle 7).

CHAPTER 20

MAKING THE FOOD PLANT INSPECTION

The most important aspect of any food plant sanitation program is the detailed food plant inspection itself. The inspection must be made daily by the food plant sanitarian and at regularly specified intervals by the food plant sanitation committee. The inspection is visible evidence to all personnel that management cares about the sanitation of the plant and operation. All employee committees and/or quality circles/teams should put sanitation on their regular meeting agenda and alert management to potential problems that they may observe. If consumer complaints should occur, the sanitarian and or the committee should thoroughly investigate the problems and take all necessary corrective actions immediately.

The food plant inspection should follow a regular routine, that is, buildings, yards and premises; followed by thorough inspection of housekeeping practices including floors, walls, ceilings, lockers, eating and rest areas; and the detailed inspection of all equipment used in the actual processes. The inspection should start with receiving and follow through to the actual shipping of the finished merchandise. The inspection should involve all areas and the inspection report should be in writing with copies to the committee members and the firm's top management. The inspection should be much more detailed than the plant survey report as suggested in Chapter 4.

A government inspection follows much the same details as does the plant inspection. When the government inspector arrives, he or she should be given the same cordial welcome that the firm extends to any other plant visitor. After presenting appropriate credentials, the firm should designate a person(s), usually the QA Manager or the Sanitarian to accompany the official inspector on a plant inspection. Prior to the plant inspection, it is highly suggested that the firm's official representative provide the official inspector a flow chart of the operation starting with incoming materials.

Every firm should always be prepared for an official government inspection. The government inspector(s) is there to make certain that the firm is in compliance with the food laws and regulations. There should be no worry by the firm, if the firm's regular inspection is in detail and all problem areas are in compliance with the rules and regulations to produce safe and sound quality foods. The government inspector may be there only to do a routine factory inspection, or make a directed inspection (concentrate on one particular aspect of the operation), or conduct a consumer complaint inspection, or follow up on a contaminated product recall inspection, and/or conduct the Good Manufacturing Practice inspection.

In any case, the inspector is entitled to enter and inspect, at reasonable times, within reasonable limits, and in a reasonable manner, establishments where foods are manufactured, processed, packed or held. They may also inspect vehicles used to transport food in interstate commerce. The word "reasonable" may cause some concern, but it has been interpreted as during normal business hours or any time a firm is in operation. The depth of the inspection is often left to the discretion of the inspector and from experience the degree of cooperation may play a significant role in defining the details of the inspection.

There are a number of suggestions that may prove wise for the firm to follow when handling a government inspection: (1) The only person that should speak with the inspector should be the firm's designated representative. (2) The firm should have firm policies relative to the use of photographic and audio equipment, types of records to be disclosed, and details of operational aspects of the process. (3) Should violations be found, the details of handling corrective actions should be declared. (4) Also, whether or not there will be a need for a search warrant. These matters should all be clarified ahead of time, but the inspector should be so notified of the firm's policy prior to making the inspection. Most importantly, if the inspector takes samples, the firm's representative should likewise take samples (I believe in taking twice the quantity that the inspector took with one-half for in-house analysis and the other half for 3rd party analysis after being notified of the violation from FDA's Analysis report). Lastly, upon completion of the inspection by the official representative, he must provide the firm with a written report (I/We observed report) if violations were found during the inspection.

Management should meet with the government representative and his designated representative to discuss any concerns or suspected violations and even though a Form 483 report is not filed, he should ask for a written statement of the inspector's findings. Further, to avoid claims and disputes that the inspector may have been denied access to given areas or information, management should ask "Did we fail to provide you with any information to which you believe you are legally entitled?" In some cases following the inspection, the inspector may follow up with an Establishment Inspection Report (EIR) setting forth in narrative form details of and about the firm along with his or her recommendations. Under the Freedom of Information Act (FOIA), a firm may request a copy of the EIR report along with other documents prepared by the inspector. These documents should be reviewed and along with the FDA Form 483 report, the firm should immediately correct any and all deviations noted on these reports as a follow up inspection may be forth coming. As stated above, if your house is in order all your products, equipment, facility, methods and personnel are in compliance, you should welcome the inspection as another pair of eyes may be helpful to know that you are doing a good job–two heads are always better than one.

Following are questions developed, in part, from those questions established by FDA and NFPA relative to food plant inspections. These questions are somewhat generic and a given processor should use these as building blocks to develop sanitation questions or inspection review for his/her particular firm and operations by the Sanitation Committee. All questions should be reviewed at each meeting of the sanitation committee. All negative responses must be explained and they should immediately be corrected by management. The food plant sanitation committee should follow the inspection process according to a previously developed flow chart for the process and note any hazards and or critical control points not in compliance.

With the availability of modern video recorders, a video of the full operation should be made at each inspection for use in the post inspection discussion by the committee and management. The sanitation committee should assign all questions that were answered negatively to designated personnel for correction. A time schedule for correcting the problem areas must be established with a follow through report of appropriate actions by the food plant sanitarian.

FIGURE 20.1 – The Inspection Must Include The Tops Of The Tanks, Pipe Lines, And General Facilities.

OUTSIDE PREMISES - ADJACENT PROPERTY, GROUNDS, BUILDINGS, AND PARKING LOTS

1. Are there potential sanitation problems around the premises or from adjacent properties?
2. Are the parking lots kept clean and free of waste?
3. Are the hedges kept trimmed and is there any trash in or around the hedges, garden areas, etc?
4. Is the trash removed from the premises daily and are the trash receptacles kept clean and closed?

RECEIVING AND SHIPPING AREAS

1. Do we have purchase specifications on all incoming materials?
2. Are all incoming materials sampled according to acceptance sampling plans and inspected before receipt?
3. Do the receipt forms indicate the supplier and the quantity delivered?
4. Are the raw materials delivered according to the specified times?
5. Is there any potential of contamination of raw materials following harvest, during shipping, or while in storage waiting for processing?
6. Are inspections made for potential contamination of raw materials?
7. What is done with products that do not meet specifications?
8. If pesticides are used on raw materials, do we have records of what was applied, when it was applied, how it was applied, and what sampling and testing is completed on raw materials before receiving them in house?
9. Are stored raw materials in-house held according to recommended storage times and temperatures for each specific item?
10. Can we identify our finished products according to given lots of raw materials?
11. Is the receiving area kept clean and in satisfactory condition?
12. Is there any evidence of insects, birds, or rodent activity in the receiving areas?

RAW MATERIAL HANDLING AND STORAGE PRACTICES

1. Is there any evidence for potential contamination of raw materials by present handling methods?
2. Are flumes, conveyors, and dumping equipment free from any molds, slime, yeast, or other microbial agents?
3. If water is used in the handling of raw materials, is the water reused? If so, what checks are made on this to be certain that it is kept clean? Is there a specified chlorine residual maintained on water used for handling raw materials?

4. If detergents are used in the flume water, is a given residual maintained and what records are kept?
5. Are emptied containers washed before returning to the field after each use?
6. Are spray washers and washing equipment maintained in good operating conditions?
7. What records are kept on washing practices, that is, water pressures, flow rates, water temperature, residual chlorine levels, and detergent residues?

HANDLING OTHER INGREDIENTS

1. Do we have purchase specifications and guarantees on all other ingredients?
2. Do we have records according to suppliers as to delivery dates, quality, and quantity?
3. Are all ingredients sampled according to acceptance sampling methods and analyzed for compliance with specifications?
4. What happens to ingredients that do not meet specifications?
5. Are all ingredients handled and stored according to acceptable conditions of time, temperature, relative humidity, light, etc.?
6. Are any rodent proofing materials, insecticides, or other pesticides used in the storage areas and what records are kept on pesticide usage?
7. Are storage areas fumigated and, if so, is the application done according to accepted practices at specified dosage levels? What records are maintained on any practices used for fumigation?
8. Are outside pest control operators used? If so, what records are kept on their activities?
9. What cleaning schedule is used for ingredient storage areas and is this maintained?
10. Do we use the practice of First In, First Out with ingredients?
11. What happens to any spilled ingredients in the warehouse?
12. Are all ingredient containers cleaned before removal of ingredients?

PRODUCT PREPARATION

1. Do we have adequate control on all preparation unit operations that have a bearing on the safety and sanitation quality of all materials in process? This includes storage time and temperature, removal of foreign substances, adequate washing and sorting, removal of any detergent residues, and the further contamination protection of all raw materials.
2. Are all unit operations checked for cleanliness before the start of shift operations?
3. Are all production rates compatible with equipment capacity?
4. Do we use in-plant chlorination? If so, what checks are made on the residuals and at what frequency?
5. Are any holding periods used in the preparation areas? If so, what controls are in place for assurance of retention of product quality?
6. What quality assurance controls are in place on each unit operation in the preparation area?
7. Do we have a clean-up schedule and procedure for each piece of equipment in the process?
8. What sampling schedule and what quality control practices are used for assuring the quality of the product during the production run?

FILLING

1. Are all empty containers accepted on the basis of acceptance sampling practices and are empty containers cleaned before filling? What records are maintained on empty containers?
2. Do we have fill specifications including temperatures for all containers by size?
3. Are all fill weight records kept and charted on a regular basis?
4. Are filled containers evaluated on a regular basis for integrity of closures?
5. Are all empty containers removed from the production line during shutdowns?
6. Are fillers and closure machines and handling equipment kept in good sanitary conditions at all times?

7. Are all containers properly coded as to a given scheme to identify commodity, type, style, size, quality, date, shift, line, plant, etc.
8. Are records maintained of filler temperature, initial temperature (IT), code, and quality per container size, style, product, by date, line, filler, etc. by products in process?
9. Are records maintained of container inspection, tear down, and integrity of closure?

PROCESSING OPERATIONS

1. Are all processing systems in compliance with "Current Good Manufacturing Practices"?
2. Are the operators certified by an approved school?
3. Are all processing times and temperatures and venting schedule posted in the processing areas for each product by style and container size?
4. Are all recorders in good working order and properly used? Are the records properly filed?
5. Are products cooled according to given schedules? Is chlorine used in the cooling water? Is there a chlorine residual maintained in the cooling water?

WAREHOUSING

1. Are proper temperatures and humidities maintained throughout the warehouse?
2. Is the warehouse maintained in according to Current Good Manufacturing Practices at all times?
3. Are doors and entry ways kept closed when not in use?
4. If air curtains, insect screens or other precautions are used, are they properly maintained and doing the job intended?
5. Are all products stored 6 inches off the floor and 18 inches away from walls?
6. Are all shipping vehicles inspected before loading and are they maintained in sanitary conditions?
7. Is there any refuse or spilled products in the warehouse or shipping areas? If so, what practice is used to remove or clean-up these areas?

CLEAN-UP PROCEDURES

1. Do we have written clean up procedures and manuals available for all cleaning personnel? Are these manuals used by cleaning personnel?
2. Are cleaning agents available to the clean-up personnel and have they been trained in how to use these chemical agents? Do they know where the MSDS sheets are and are these available to authorized users?
3. Does the cleaning supervisor have check lists for the evaluation of adequacy of cleaning each unit operations and the total facility? Who audits this check list?
4. Do cleaning personnel have sufficient time and training to keep the plant clean?

Every firm should develop questions that are suitable to their operation and they should use these to make their own inspection rather than wait for an official inspection that could prove costly. Inspections are a must in the food industry and they must be accomplished on a regularly scheduled basis with complete and thorough follow through to assure that the plant is clean and safe at all times.

REFERENCES

Anon. Do Your Own Establishment Inspection. U. S. Department Of Health and Human Services. HHS Publi. No. (FDA) 82-2163.
Bryan, Frank L. 1974 Microbiological Food Hazards today-Based on Epidemiological Information. Food Technology Sept. 59-84.
Corlett, Donald A. Jr. 1989. Refrigerated Foods and Use of Hazard Analysis and Critical Control Point Principles. Food Technology. February 91-94.
Gould, Wilbur A. 1980. Good Manufacturing Practices for Snack Food Manufacturers. Snack Food Association, Alexandria, VA.
Ito, Keith. 1974. Microbiological Critical Control Points in Canned Foods. Food Technology Sept. 46-47.
Kauffman, F. Leo. 1974 How FDA Uses HACCP. Food Technology Sept. 51-52.
Peterson, A. C. and R. E. Gunnerson. Microbiological Critical Control Points in Frozen Foods. Food Technology Sept. 37-44.
Somers, Ira I. 1973 FDA-HACCP Inspection, Suggestions for the Canner. NCA Bull. 35 L. Washington, D. C.

CHAPTER 21

RECALLS

Dr. Alexander M. Schmidt in 1973, the then Commissioner of the Food and Drug Administration, stated that "An important consumer protection responsibility for the Food and Drug Administration is to seek out and to assure prompt removal of defective or hazardous products from the market...Recall usually is a more efficient and practical means for reversing the chain of product distribution. Especially is recall preferable to seizure when the defective product is dangerous to health and speed in retrieval is all important" This new procedure has saved the industry much embarrassment and by the cooperation of the industry with FDA great savings in dollars.

Simply put, the recall or product withdrawal requires the location of the defective or adulterated product that is in the marketing channels, its removal from the marketing channels, and the accumulation and filing of accurate information with the FDA. The management of food firms has the responsibility to manufacture and market safe and wholesome products. This is both a legal and a moral responsibility. If defective products are manufactured or in the market place, the firm bears the full responsibilities to remove these products promptly and effectively. This too is a moral and legal responsibility.

The following are the Recall Classifications and normal FDA Actions:

CLASS I - This is an emergency situation involving removal from the market of products in which the consequences are immediate or long-term, life-threatening and involve a direct cause-effect relationship, that is, *Clostridium botulinum* in foods. The normal FDA action is to make a product recall to the consumer level as complete and as rapidly as possible with 100% effectiveness check of known direct and sub-distribution points including the consumer. The FDA places a Class I recall on the public recall list and may issue a public warning.

CLASS II - This is a priority situation in which the consequences may be immediate or long-range and possible or potentially life threatening or hazardous to health, that is, pathogenic organisms in food exclusive of *Clostridium botulinum.* The normal FDA action is a product recall to retail level completely and promptly with an effectiveness check on a sliding scale depending on the seriousness of the hazard, ranging upward from 2 - 10% of known direct distribution points and upward from 1 - 2 sub-distribution points for each direct distribution point checked. It may be placed on the public recall list as a Class II recall and there is a possibility of the issuance of press releases as the circumstances may warrant.

CLASS III- This is a routine situation in which the consequences to life (if any) are remote or non-existent. The products are recalled because of adulteration (filth in food relating to aesthetics quality) or misbranding (label violations) and the product does not involve a health hazard. The normal FDA action is to make the product recall to the wholesale level with no effectiveness checks. Generally, Class III recalls are placed on public recall list as a Class III Recall. Further, FDA does not ordinarily make a press release, but they will respond to inquiries from the press and the public.

Product withdrawals are of two types, that is Type A which is a situation where none of the product has left the direct control of the manufacture or primary distributor. FDA checks on the adequacy of the withdrawal as for appropriate calls of recall. They do not place the product on the public recall list nor do they initiate a press release, but FDA will respond to inquires from the press and the public. A Type B withdrawal is a situation where the product is not in violation or if the product is in violation, the violations are minor and not subject to seizure under current FDA policies or guidelines and FDA does not take any action.

Every food firm should have a formal RECALL ORGANIZATION. The group should have a Chairman and in many food firms, the Chairman and coordinator is the Quality Assurance Manager. Working with him are the Production Manager, the Firm's financial officer, the Marketing Manager, and the firm's

Legal Counsel. Alternates should be appointed for each of the above and have full authority to act in their absence.

The duties of this recall group are to meet immediately at the request of the Recall Coordinator and devote all the time necessary to plan and implement a recall if a recall is called for. In the case of a firm initiated recall, they access the seriousness of the hazard and determine the depth of the recall required or the withdrawal of any product involved. If the recall is at the request of FDA, the group reviews the findings of FDA and they reach an agreement on the recall and the depth of the recall required. The Company President must be kept informed of the findings in all cases.

The following information is required by FDA for a firm initiated recall:

→ Identity of product involved
→ Reason for the recall
→ Evaluation of the risk
→ Total amount of product involved and the time period covered
→ Amount of product outside of firm's control
→ Number and possible identity of direct accounts
→ A copy of any recall communications
→ Proposed recall strategy
→ Name of Company official to contact.

As the past has clearly indicated, Recalls can be most devastating. Recalls should never happen and they will not happen if a firm has established a complete and thorough sanitation program with effectiveness checks on all incoming materials, products in process and all finished products. However, recalls have happened in the past and food firms need to be prepared to act and to act swiftly. A food firm must have a Recall Group and they must know what to do. Advance planning and mock recall trials are essential for the good of the firm. All actions must be prompt, accurate, and complete with thorough communications with FDA.

The details of the Recall authority and procedures as elucidated by FDA follow:

Subpart B - (Reserved)

Subpart C - Recalls (Including Product Corrections) Guidelines on Policy, Procedures, and Industry Responsibilities

AUTHORITY: Sec. 1 et seq., Pub. L. 717, 52 Stat. 1040-1059 as amended (21 U.S.C. 301 et seq.); secs. 301, 351, and 361, Pub. L 410, 58 Stat. 691-703 as amended (42 U.S.C. 241, 262, and 264).
SOURCE: 43 FR 26218, June 16, 1978, unless otherwise noted.

§ 7.40 Recall policy.

(a) Recall is an effective method of removing or correcting consumer products that are in violation of laws administered by the Food and Drug Administration. Recall is a voluntary action that takes place because manufacturers and distributors carry out their responsibility to protect the public health and well-being from products that present a risk of injury or gross deception or are otherwise defective. This section and §7.41 through 7.59 recognize the voluntary nature of recall by providing guidelines so that responsible firms may effectively discharge their recall responsibilities. These sections also recognize that recall is an alternative to a Food and Drug Administration initiated court action for removing or correcting violative, distributed products by setting forth specific recall procedures for the Food and Drug Administration to monitor recalls and assess the adequacy of a firm's efforts in recall.

(b) Recall may be undertaken voluntarily and at any time by manufacturers and distributors, or at the request of the Food and Drug Administration. A request by the Food and Drug Administration that a firm recall a product is reserved for urgent situations and is to be directed to the firm that has primary responsibility for the manufacture and marketing of the product that is to be recalled.

(c) Recall is generally more appropriate and affords better protection for consumers than seizure, when many lots of product have been widely distributed. Seizure, multiple seizure, or other court action is indicated when a firm refuses to undertake a recall requested by the Food and Drug Administration, or where the agency has reason to believe that a recall would not be effective, determines that a recall is ineffective, or discovers that a violation is continuing.

§ 7.41 Health hazard evaluation and recall classification.

(a) An evaluation of the health hazard presented by a product being recalled or considered for recall will be conducted by an ad hoc committee of Food and Drug Administration scientists and will take into account, but need not be limited to, the following factors:

(1) Whether any disease or injuries have already occurred from the use of the product.
(2) Whether any existing conditions could contribute to a clinical situation that could expose humans or animals to a health hazard. Any conclusion shall be supported as completely as possible by scientific documentation and/or statements that the conclusion is the opinion of the individual(s) making the health hazard determination.
(3) Assessment of hazard to various segments of the population, e.g., children, surgical patients, pets, livestock, etc., who are expected to be exposed to the product being considered, with particular attention paid to the hazard to those individuals who may be at greatest risk.
(4) Assessment of the degree of seriousness of the health hazard to which the populations at risk would be exposed.
(5) Assessment of the likelihood of occurrence of the hazard.
(6) Assessment of the consequences (immediate or long-range) of occurrence of the hazard.

(b) On the basis of this determination, the Food and Drug Administration will assign the recall a classification, i.e., Class I, Class II, or Class III, to indicate the relative degree of health hazard of the product being recalled or considered for recall.

§ Recall strategy.

(a) *General.*

(1) A recall strategy that takes into account the following factors will be developed by the agency for a Food and Drug Administration-requested recall and by the recalling firm for a firm-initiated recall to suit the individual circumstances of the particular recall:
(i) Results of health hazard evaluation.
(ii) Ease in identifying the product.
(iii) Degree to which the product's deficiency is obvious to the consumer or user.
(iv) Degree to which the product remains unused in the market-place.
(v) Continued availability of essential products.

(2) The Food and Drug Administration will review the adequacy of a proposed recall strategy developed by a recalling firm and recommend changes as appropriate. A recalling firm should conduct the recall in accordance with an approved recall strategy but need not delay initiation of a recall pending review of its recall strategy.

(b) **Elements of a recall strategy**. A recall strategy will address the following elements regarding the conduct of the recall:

(1) **Depth of recall**. Depending on the product's degree of hazard and extent of distribution, the recall strategy will specify the level in the distribution chain to which the recall is to extend, as follows:

(i) consumer or user level, which may vary with product, including any intermediate wholesale or retail level; or

(ii) Retail level, including any intermediate wholesale level; or

(iii) Wholesale level.

(2) **Public warning**. The purpose of a public warning is to alert the public that a product being recalled presents a serious hazard to health. It is reserved for urgent situations where other means for preventing use of the recalled product appear inadequate. The Food and Drug Administration in consultation with the recalling firm will ordinarily issue such publicity. The recalling firm that decides to issue its own public warning is requested to submit its proposed public warning and plan for distribution of the warning for review and comment by the Food and Drug Administration. The recall strategy will specify whether a public warning is needed and whether it will issue as:

(i) General public warning through the general news media, either national or local as appropriate, or

(ii) Public warning through specialized news media, e.g., professional or trade press, or to specific segments of the population such as physicians, hospitals, etc.

(3) **Effectiveness checks**. The purpose of effectiveness checks is to verify that all consignees at the recall depth specified by the strategy have received notification about the recall and have taken appropriate action. The method for contacting consignees may be accomplished by personal visits, telephone calls, letters, or a combination thereof. A guide entitled "Methods for Conducting Recall Effectiveness Checks" that describes the use of these different methods is available upon request from the Dockets Management Branch (HFA-305), Food and Drug Administration, Room 4-62, 5600 Fishers Lane, Rockville, MD 20857. The recalling firm will ordinarily be responsible for conducting effectiveness checks, but the Food and Drug Administration will assist in this task where necessary and appropriate. The recall strategy will specify the method(s) to be used for and the level of effectiveness checks that will be conducted, as follows:

(i) Level A–100 percent of the total number of consignees to be contacted;

(ii) Level B–Some percentage of the total number of consignees to be contacted, which percentage is to be determined on

a case-by-case basis, but is greater than 10 percent and less than 100 percent of the total number of consignees;
(iii) Level C–10 percent of the total number of consignees to be contacted;
(iv) Level D–2 percent of the total number of consignees to be contacted; or
(v) Level E–No effectiveness checks.

[43 FR 26218, June 16, 1978, as amended at 46 FR 8455, Jan. 27, 1981]

§ 7.45 Food and Drug Administration - requested recall.

(a) The Commissioner of Food and Drugs or his designee under §5.20 of this chapter may request a firm to initiate a recall when the following determinations have been made:

(1) That a product that has been distributed presents a risk of illness or injury or gross consumer deception.
(2) That the firm has not initiated a recall of the product.
(3) Than an agency action is necessary to protect the public health and welfare.

(b) The Commissioner or his disignee will notify the firm of this determination and of the need to begin immediately a recall of the product. Such notification will be by letter or telegram to a responsible official of the firm, but may be preceded by oral communication or by a visit from an authorized representative of the local Food and Drug Administration district office, with formal, written confirmation from the Commissioner or his designee afterward. The notification will specify the violation, the health hazard classification of the violative product, the recall strategy, and other appropriate instructions for conducting the recall.

(c) Upon receipt of a request to recall, the firm may be asked to provide the Food and Drug Administration any or all of the information listed in § 7.46(a). The firm, upon agreeing to the recall request, may also provide other information relevant to the agency's determination of the need for the recall or how the recall should be conducted.

§ 7.46 Firm-initiated recall.

(a) A firm may decide of its own violation and under any circumstances to remove or correct a distributed product. A firm that does so because it believes the product to be violative is requested to notify immediately the appropriate Food and Drug Administration district office listed in §5.115 of this chapter. Such removal or correction will be considered a recall only if the Food and Drug Administration regards the product as involving a violation that is subject to legal action, e.g., seizure. In such cases, the firm will be asked to provide the Food and Drug Administration the following information:

(1) Identity of the product involved.
(2) Reason for the removal or correction and the date and circumstances under which the product deficiency or possible deficiency was discovered.
(3) Evaluation of the risk associated with the deficiency or possible deficiency.
(4) Total amount of such products produced and/or the timespan of the production.
(5) Total amount of such products estimated to be in distribution channels.
(6) Distribution information, including the number of direct accounts and, where necessary, the identity of the direct accounts.
(7) A copy of the firm's recall communication if any has issued, or a proposed communication if none has issued.
(8) Proposed strategy for conducting the recall.
(9) Name and telephone number of the firm official who should be contacted concerning the recall.

(b) The Food and Drug Administration will review the information submitted, advise the firm of the assigned recall classification, recommend any appropriate changes in the firm's strategy for the recall, and advise the firm that its recall will be placed in the weekly FDA Enforcement Report. Pending this review, the firm need not delay initiation of its product removal or correction.

(c) A firm may decide to recall a product when informed by the Food and Drug Administration that the agency has determined that the product in question violates the law, but the agency has not specifically requested a recall. The firm's action also is considered a firm-initiated recall and is subject to paragraphs (a) and (b) of this section.

(d) A firm that initiates a removal or correction of its product which the firm believes is a market withdrawal should consult with the appropriate Food and Drug Administration district office when the reason for the removal or correction is not obvious or clearly understood but where it is apparent, e.g., because of complaints or adverse reactions regarding the product, that the product is deficient in some respect. In such cases, the Food and Drug Administration will assist the firm in determining the exact nature of the problem.

§ 7.49 Recall communications.

(a) *General.* A recalling firm is responsible for promptly notifying each of its affected direct accounts about the recall. The format, content, and extent of a recall communication should be commensurate with the hazard of the product being recalled and the strategy developed for that recall. In general terms, the purpose of a recall communication is to convey:

(1) That the product in question is subject to a recall.
(2) That further distribution or use of any remaining product should cease immediately.
(3) Where appropriate, that the direct account should in turn notify its customers who received the product about the recall.
(4) Instructions regarding what to do with the product.

(b) *Implementation.* A recall communication can be accomplished by telegrams, mailgrams, or first class letters conspicuously marked, preferably in bold red type, on the letter and the envelope: "DRUG [or FOOD, BIOLOGIC, etc.] RECALL [or CORRECTION"]. The letter and the envelope should be also marked: "URGENT" for class I and class II recalls and, when appropriate, for class III recalls. Telephone calls or other personal contacts should ordinarily be confirmed by one of the above methods and/or documented in an appropriate manner.

(c) *Contents.*

(1) A recall communication should be written in accordance with the following guidelines:

(i) Be brief and to the point.
(ii) Identify clearly the product, size, lot number(s), code(s) or serial number(s), and any other pertinent descriptive information to enable accurate and immediate identification of the product;
(iii) Explain concisely the reason for the recall and the hazard involved, if any;
(iv) Provide specific instructions on what should be done with respect to the recalled products; and
(v) Provide a ready means for the recipient of the communication to report to the recalling firm whether it has any of the product, e.g., by sending a postage-paid, self-addressed postcard or by allowing the recipient to place a collect call to the recalling firm.

(2) The recall communication should not contain irrelevant qualifications, promotional materials, or any other statement that may detract from the message. Where necessary, followup communications should be sent to those who fail to respond to the initial recall communication.

(d) *Responsibility* of recipient. Consignees that receive a recall communication should immediately carry out the instructions set forth by the recalling firm and, where necessary, extend the recall to its consignees in accordance with paragraphs (b) and (c) of this section.

§ 7.50 Public notification of recall.

The Food and Drug Administration will promptly make available to the public in the weekly FDA Enforcement Report a descriptive listing of each new recall according to its classification, whether it was Food and Drug

Administration-requested or firm-initiated, and the specific action being taken by the recalling firm. The Food and Drug Administration will intentionally delay public notification of recalls of certain drugs and devices where the agency determines that public notification may cause unnecessary and harmful anxiety in patients and that initial consultation between patients and their physicians is essential. The report will not include a firm's product removals or corrections which the agency determines to be market withdrawals or stock recoveries. The report, which also includes other food and Drug Administration regulatory actions, e.g., seizures that were effected and injunctions and prosecutions that were filed, is available upon request from the Office of Public Affairs (HFI-1), Food and Drug Administration, 5600 Fishers Lane, Rockville, MD 20857.

§ 7.54 Recall status reports.

(a) The recalling firm is requested to submit periodic recall status reports to the appropriate Food and Drug Administration district office so that the agency may assess the progress of the recall. The frequency of such reports will be determined by the relative urgency of the recall and will be specified by the Food and Drug Administration in each recall case; generally the reporting interval will be between 2 and 4 weeks.

(b) Unless otherwise specified or inappropriate in a given recall case, the recall status report should contain the following information:

(1) Number of consignees notified of the recall, and date and method of notification.

(2) Number of consignees responding to the recall communication and quantity of products on hand at the time it was received.

(3) Number of consignees that did not respond (if needed, the identity of nonresponding consignees may be requested by the Food and Drug Administration).

(4) Number of products returned or corrected by each consignee contacted and the quantity of products accounted for.

(5) Number and results of effectiveness checks that were made.

(6) Estimated time frames for completion of the recall.

(c) Recall status reports are to be discontinued when the recall is terminated by the Food and Drug Administration.

§ 7.55 Termination of a recall.

(a) A recall will be terminated when the Food and Drug Administration determines that all reasonable efforts have been made to remove or correct the product in accordance with the recall strategy, and when it is reasonable to assume that the product subject to the recall has been removed and proper disposition or correction has been made commensurate with the degree of hazard of the recalled product. Written notification that a recall is terminated will be issued by the appropriate Food and Drug Administration district office to the recalling firm.

(b) A recalling firm may request termination of its recall by submitting a written request to the appropriate Food and Drug Administration district office stating that the recall is effective in accordance with the criteria set forth in paragraph (a) of this section, and by accompanying the request with the most current recall status report and a description of the disposition of the recalled product.

§ 7.59 General Industry guidance.

A recall can be disruptive of a firm's operation and business, but there are several steps a prudent firm can take in advance to minimize this disruptive effect. Notwithstanding similar specific requirements for certain products in other parts of this chapter, the following is provided by the Food and Drug Administration as guidance for a firm's consideration:

(a) Prepare and maintain a current written contingency plan for use in initiating and effecting a recall in accordance with §7.40 through 7.49, 7.53, and 7.55.

(b) Use sufficient coding of regulated products to make possible positive lot identification and to facilitate effective recall of all violative lots.

(c) Maintain such product distribution records as are necessary to facilitate location of products that are being recalled. Such records should be maintained for a period of time that exceeds the shelf life and expected use of the product and is at least the length of time specified in other applicable regulations concerning records retention.

Subpart D - Infant Formula Recalls

AUTHORITY: Secs. 412, 701(a), 52 Stat. 1055, 94 Stat. 1190-1192 (21 U.S.C. 350a, 371(a)).

SOURCE: 47 FR 18835, Apr. 30, 1982, unless otherwise noted.

§ 7.70 Scope and effect.

(a) The criteria in this subpart apply to a recall of an infant formula initiated by a manufacturer under section 412(d) of the act. The requirements of this subpart apply only when a manufacturer has determined to remove from the market an infant formula that has been distributed, that is no longer subject to the control of the manufacturer, and that is in violation of the laws and regulations administered by the Food and Drug Administration and against which the agency could initiate legal or regulatory action. The Food and Drug Administration will monitor continually the recall action and will take appropriate actions to ensure that the violative infant formula is removed from the marketplace.

(b) The failure of a recalling firm to comply with the regulations of this subpart is a prohibited act under section 301(s) of the act.

§ 7.71 Elements of an infant formula recall.

A recalling firm shall conduct an infant formula recall with the following elements:

(a) The recalling firm shall evaluate in writing the hazard to human health associated with the use of the infant formula. This health hazard evaluation shall include consideration of any disease, injury, or other adverse physiological effect that has been or that could be caused by the infant formula and of the seriousness, likelihood, and consequences of the disease, injury, or other adverse physiological effect.

(b) The recalling firm shall devise a written recall strategy suited to the individual circumstances of the particular recall. The recall strategy shall take into account the health hazard evaluation and specify the following: the depth of the recall; if necessary, the public warning to be given about any hazard presented by the infant formula; the disposition of the recalled infant formula; and the effectiveness checks that will be made to determine that the recall is carried out.

(c) The recalling firm shall promptly notify each of its affected direct accounts about the recall. The format of a recall communication shall be distinctive and the content and extent of a recall communication shall be commensurate with the hazard of the infant formula being recalled and the strategy developed for the recall. The recall communication shall instruct consignees to report back quickly to the recalling firm about whether they are in possession of the recalled infant formula and shall include a means of doing so. The recall communication shall also advise consignees of how to return the recalled infant formula to the manufacturer or otherwise dispose of it. The recalling firm shall send a followup recall communication to any consignee that does not respond to the initial recall communication.

(d) The recalling firm shall furnish promptly to the appropriate Food and Drug Administration district office listed in §5.115 of this chapter, as they are available, copies of the health hazard evaluation, the recall strategy, and all recall communications directed to consignees, distributors, retailers, and members of the public.

§ 7.72 Reports about an infant formula recall.

(a) *Telephone report.* When a determination is made that an infant formula is to be recalled, the recalling firm shall promptly telephone the appropriate Food and Drug Administration district office listed in §5.115 of this chapter and shall provide relevant information about the infant formula that is to be recalled.

(b) *Initial written report.* Within 14 days after the recall has begun, the recalling firm shall provide a written report to the appropriate Food and Drug Administration district office. The report shall contain relevant information, including the following cumulative information concerning the infant formula that is being recalled:

(1) Number of consignees notified of the recall, and date and method of notification.
(2) Number of consignees responding to the recall communication and quantity of recalled infant formula on hand at the time it was received.
(3) Quantity of recalled infant formula returned or corrected by each consignee contacted and the quantity of recalled infant formula accounted for.
(4) Number and results of effectiveness checks that were made.
(5) Estimated time frames for completion of the recall.

(c) *Status reports.* The recalling firm shall submit to the appropriate Food and Drug Administration district office a written status report on the recall at least every 14 days until the recall is terminated. The status report shall describe the steps taken by the recalling firm to carry out the recall since the last report and the results of those steps.

§ 7.73 Termination of recall.

The recalling firm may submit a recommendation for termination of the recall to the appropriate Food and Drug Administration district office listed in §5.115 of this chapter for transmittal to the Division of Regulatory Guidance, Bureau of Foods, for action. Any such recommendation shall contain information supporting a conclusion that the recall strategy has been effective. The agency will respond within 15 days of receipt by the Division of Regulatory Guidance, Bureau of Foods, of the request for termination. The recalling firm shall continue to implement the recall strategy until it receives final written notification from the agency that the recall has been terminated. The agency will send such a notification unless it has information, from FDA's own audits or from other sources, demonstrating that the recall has not been effective. The agency may conclude that a recall has not been effective if:

(a) The recalling firm's distributors have failed to retrieve the recalled infant formula.

(b) Stocks of the recalled infant formula remain at the distribution level from which the infant formula was recalled.

§ 7.74 Revision of recall.

If after a review of the recalling firm's recall strategy, health hazard evaluation, or periodic reports or other monitoring of the recall, the Food and Drug Administration concludes that the actions of the recalling firm are deficient, the agency shall notify the recalling firm of any serious deficiency. The agency may require the firm to:

(a) Change the depth of the recall, if the agency believes on the basis of the available scientific data that the health hazard of the recalled infant formula is greater than that described in the health hazard evaluation, or if the agency concludes that the depth of recall is not adequate in light of the health hazard evaluation submitted by the firm.

(b) Carry out additional effectiveness checks, if the agency's audits, or other information, demonstrate that the recall has not been effective.

(c) Issue additional notifications to the firm's direct accounts, if the agency's audits, or other information, demonstrate that the original notifications were not received, or were disregarded, in a significant number of cases.

§ 7.75 Compliance with this subpart.

A recalling firm may satisfy the requirements of this subpart by any means reasonably calculated to meet the obligations set forth above. The recall guidelines in Subpart C of Part 7 specify procedures that may be useful to a recalling firm in determining how to comply with these regulations.

CHAPTER 22

THE ROLE OF QUALITY ASSURANCE IN FOOD PLANT SANITATION

Quality assurance personnel are generally assigned the responsibility of food plant inspection and overall control of food plant sanitation in a food plant. These are duties beyond the evaluation of incoming materials for compliance to specifications, for product control during the processing, and for the actual audit of all finished products for conformance to requirements. These responsibilities are of major significance in the total quality assurance program of a food company.

Part 110.80 of the CGMP's covers quality control operations for food and food packaging operations. The rule requires food firms to adhere to adequate sanitation principles in all of the production procedures, including receiving, inspecting, transporting, preparing, manufacturing, packaging, and storage of foods. Someone must be assigned this responsibility and it usually falls to the Quality Assurance Manager or his Associates. The general rule is that "all reasonable precautions shall be taken to ensure that production procedures do not contribute contamination from any source".

To meet this rule, food firms must periodically evaluate their products for chemicals, microbial, or extraneous material contamination. If any food material is found contaminated, the manufacturer must determine whether the contamination renders the food "adulterated" within the meaning of the FDC Act. If the food is adulterated, it must either be rejected or reprocessed to eliminate any contamination. A contaminated batch may not be blended with uncontaminated batches, even if the final product would not be excessively contaminated.

All raw materials must be inspected to ensure that they are clean and suitable for processing, and must be properly stored to avoid contamination and minimize deterioration. Part 110.80 (a) states that water used for washing, rinsing, or conveying food shall be safe and of adequate sanitary quality. The water may be reused for washing, rinsing, or conveying food if it does not increase the level of contamination of the food.

Part 110.80 (b) primarily details the means by which a manufacturer may prevent or minimize the growth of unwanted microorganisms in food. The rule recommends that manufacturers monitor such factors as time, temperature, humidity, a sub W, pressure, flow rate, and pH to ensure that mechanical breakdowns, delays, and temperature fluctuations do not contribute to the decomposition or contamination of food. Details relative process conditions are also spelled out in this section of the rule.

Most importantly, the rule, part 110.80 (13) (i) states "use of quality control operation in which the critical control points are identified and controlled during manufacture". This is what modern Total Quality Assurance is all about.

Total Quality Assurance is the modern term used by many food firms to assure management of the quality of incoming materials, materials in process, and the qualities of all finished products as well as products in the marketing channel. Total quality assurance is an umbrella term to cover the major aspects of quality assurance, that is, control, evaluation, audit, and activities such as R&D, food plant sanitation, and trouble shooting.

The quality assurance department's primary function is to provide confidence for management and the ultimate consumer that the product and processes are what the firm has specified. A firm is in business to produce products--intended for sale to customers--from which the firm hopes to make a profit. The key word is the customer. The customer is the one a firm must satisfy and it is the customer who ultimately establishes the level of quality the firm manufactures. The customer is management's guide to quality and this is what the firm builds its specifications around and the label requirements. The firm must keep the customer satisfied if they expect to build repeat business.

Quality assurance has the responsibilities to see that the customer is satisfied. Quality assurance does this by building a control program and regulating all processes to given standards of performance. These standards become very specific for each unit operation in the process.

FIGURE 22.1 - Quality Assurance Personnel Establish The Paremeters Of Operation With The Line Operator

A second major responsibility of the quality assurance program is to evaluate the entire operation. That is, they must appraise the worth of all incoming materials, materials in process, and all finished products. Appraisal includes both subjective or sensory evaluation and objective evaluation or physical, chemical and microbiological testing for the various characteristics or attributes of the product, processes, personnel, equipment, facilities, etc.

A third responsibilities of the quality assurance department is the actual audit of the firm's product in the warehouse, in the marketing channels, competing products, and the firm's product during its shelf life. Auditing includes compliance with label requirements, nutritional data in some cases, safety, and overall product quality. Techniques are similar to those used in product evaluation, that is, subjective and objective measurements.

A fourth responsibility of the quality assurance department may include trouble shooting, research and development, waste disposal, and the many details of food plant inspection and food plant sanitation. Thus, the quality assurance personnel provide much valuable information and assistance to the success of any food firm. They have been aptly called the nerve center of the food operation as they do provide a constant flow of information to keep the firm and its personnel "in the know" and "on the grow".

Quality assurance personnel generally have training in chemistry, microbiology, engineering, as well as processing and technology pertaining to specific commodities. Graduate level quality assurance personnel may have extensions of the above basic areas and they should have a thorough understanding of research, that is, how to organize and conduct scientific studies to solve problems. Most graduate quality assurance personnel should be thoroughly knowledgeable about sanitation and the CGMP's. All quality assurance personnel must be able to communicate well, that is, they must be able to write, speak, and visualize solutions to problems and communicate this information to personnel so that they can comprehend the what, how and why of the situation.

Quality assurance personnel should be part of the management team. They should be responsible for determining the process capabilities of the entire operation by each individual unit operation, they should be capable of planning the total quality assurance program, that is, the control, evaluation and audit of the process operations and all the products from receiving to warehousing. They should evaluate all raw ingredients and foods in process and be capable of handling complaint data and pinpoint it to time of production from their records and take corrective actions. They must be able to generate the cost of quality and the price of non-conformance. Further quality assurance personnel must provide feed back information to management and all production personnel. Quality assurance personnel must know how to solve problems through the use of cause and effect diagrams, process flow diagrams, statistical techniques, control charts, and Pareto principles. They must be able to identify problems, determine the cause, solutions for the causes, and how to control the process.

Quality assurance personnel must know the latest in quick methods for determining product quality and how to conduct inspections to eliminate hazards in the plant. They must know the critical control points of the entire operation and they must be capable of controlling all critical control points to eliminate any potential problems in the process or of the products.

Quality assurance personnel must be capable of carrying all of the above and the food plant inspections and they must be the spokes person on all sanitation and quality problems for the firm. They have major responsibilities in the success of any food firm.

REFERENCE

Gould, Wilbur A. and Gould, Ronald W. 1988 Total Quality Assurance for the Food Industries. CTI Publications, Inc. Baltimore, Md.

CHAPTER 23

SANITATION TRAINING OF THE FOOD PLANT WORKER

Training is teaching the worker how to do the job or task properly. It is giving of instruction so as to make the worker proficient and qualified. Training is the pathway to growth, confidence, self esteem, belief in self, and independence.

If an organization wishes to improve its position in the industry, it must focus on people. People make a company. Decisions and plans are made by people. Work is done by people. In reality, continuous improvement will only be achieved through people. Therefore, the growth of a firm is based on the growth of its people.

The overall objectives of training are (1) to support the needs of the food firm, (2) to teach new skills, (3) to develop proper attitude of all workers necessary to follow prescribed policies and procedures, and (4) to convey new knowledge to assure the manufacture of safe and wholesome foods.

Good sanitation in a food plant through proper training of all the workers should have a positive return on the investment. Employee moral is an important benefit of sanitation training. People will develop pride in their work once they understand the importance of food plant sanitation.

There are four basic essentials to any training program. These are:

(1) **The plan.** What we want our people to understand and become proficient about. In other words, What is to be taught; Why is the training necessary; How will the training be done; Who will do the training; Where will the training be given; and When will the training take place.? Planning properly will make the investment pay big returns.

Every firm is different and the plan should be adapted to the given firm. Although basic sanitation fundamentals are the same for most firms, there are differences due to commodities being processed by given firms. Every plan should include an understanding of microorganisms, the importance of personal hygiene, how and what to clean, and a thorough understanding of the potential hazards and how to develop and control the critical control points. The plan should include a given sanitation text to be used during the course of study and other handout materials for all attendees. It should include the use of available visual aides, posters, videos and movies, charts and graphs, and specific cases of insanitary practices as reported by FDA and other agencies. The plan should also include methods for evaluating the trainees following the course of instruction.

(2) **Method of Instruction**. This will depend in great part on the number of people to be trained. It may be possible to give the training on the job through the use of one-on-one or the Big Brother approach. I do not like the latter as some times big brother is not thorough and the trainee does not get the big picture. One on one is a satisfactory method, but the training should be done by the supervisor or management, never the big brother approach.

If several are to be trained, the trainees should attend formal classes, workshops, or seminars with competent instructor(s). The use of consultants, suppliers, and management is very effective. Regardless of method of instruction, the instruction must be interesting and full of examples that are relevant. The emphasis should be on helping the worker to understand how to operate his or her equipment or work station in a safe and sanitary manner. The worker must understand that contamination of any kind is not tolerable, that food spoilage is never an accident, that food borne illnesses are the result of inferior raw materials or inadequate or inappropriate procedures in food handling and manufacture, and that continued improvement of the quality of the finished product is their direct responsibility.

The most important aspect of the instruction is the actual instructor. This person must be enthusiastic, knowledgeable, understand the firms needs, capable of speaking fluently and with ease, and have a genuine interest in the subject. Teaching is not easy and not everyone is a capable instructor. The whole intent of the training should be to help the worker feel confident in the area of food plant sanitation and the manufacture of clean, safe, and wholesome foods. Obviously, a secondary benefit of the training is to make the worker feel secure about their job and develop an improved sense of pride about their work.

Ideally, the training should be off-site in a neutral facility where the attendee can feel at ease during the lecture-discussion portions of the training. Training should include actual demonstrations at the firm's factory by the instructor and key personnel. However, I much prefer to give each "student" the hands on approach. This probably should also be at the plant site where the worker learns to work with the firm's sanitation equipment, chemicals, and procedures. This so-called hands on approach is like a laboratory experiment. The "student" follows given procedures, generates outcomes or data, and then evaluates the results. Good instructors should have the "student" interpret the results by drawing conclusions and making recommendations. This laboratory practice is without question one of the most ideal ways to train personnel. Many instructors do not like to follow this approach as it requires a large amount of preparation time to set-up each experiment, additional resources, and too much time and effort to properly evaluate each trainee's laboratory report.

(3) **Checking**. Regardless of the method of instruction, the worker must be evaluated to show the effectiveness of the training. The evaluation may be one-on-one discussions, oral or written questions, or individual presentation before the class. I prefer using all of the above, but because of time and, if a large number of trainees, I use written questions with the trainee providing the answers. The questions for the worker should be in the form of multiple choice, true or false statements, matching, or fill in blank spaces. Regardless of the type of examination the individual trainee should be given the results of the examination promptly and, hopefully, praised for their efforts, thus reinforcing them with a positive attitude about their job.

(4) **Act**. Here the worker should actually get to work and act out what has been taught. Obviously, if the worker does not perform as expected, he or she may have to take the training over again. There is one constant rule that must be remembered, that is, if the worker doesn't understand, they cannot perform as

expected. There is a Chinese proverb that I like to use that says a whole lot about training and the effectiveness of the trainee and his or her understanding:

"Tell me and I will forget
Show me and I may remember
Involve me and I will understand"

Training is not easy and it is not the whole answer, but it is the first step in assuring that your people understand the need for, the what of, the how to, the why of, etc. of food plant sanitation and the safety of the food with which they work.

Obviously, food firms should set aside resources for training personnel. These resources will vary with the depth of the training and how the training is conducted. The resources really are a minute amount when one considers the potential cost of producing food that is not in compliance with the customer's expectations. Some food firms set aside 2 to 5% of their operating capitol for training and educational programs to help their people grow to keep up with the growth of the firm.

The workers that are to be trained must remove the fear of loss of their jobs and they must want to listen, observe and study to learn. Workers that are not interested in listening, observing new practices, or unwilling to learn should find other employment. The outcome of training will be continuous improvement for the firm. Further, the firm will heap rewards in the eye of the user as the user will always be able to purchase what they expect.

Training is a never ending subject. New changes, information, help, and procedures are constantly being brought to the fore front. It behooves management and, in particular, the sanitarian to stay up-to-date and constantly re-train the worker(s).

Some of the ways to stay up-to-date include:

(1) Reading the Federal Register and studying the Code of Federal Regulations,

(2) Staying alert to new books published on the subject,

(3) Reading the Trade and Scientific literature that deal with food plant sanitation,

(4) Working with your food plant sanitation supplier, your equipment supplier, your material supplier, and technical representatives of your chemical supplier(s),

(5) Attending meetings, seminars, workshops and listening, talking, and consulting with your peers, and

(6) Joining a sanitation organization and becoming active in how to solve sanitation problems with your fellow members.

Safe foods are never an accident. They are the result of intelligent management and the trained worker working together with one common goal, that is, the production of safe, wholesome foods that far exceed any government standard. Safe foods are the outcome of understanding what is expected at all times.

APPENDIX

FEDERAL FOOD, DRUG, AND COSMETIC ACT AS AMENDED

CONTENTS

(*References in brackets [ ] are to title 21 U.S. Code)

Sec. 404. Emergency permit control [344]

Sec. 405. Regulations making exemptions [345]
Sec. 406. Tolerances for poisonous ingredients in food [346]
Sec. 407. Oleomargarine or margarine [347]
Sec. 408. Tolerances for pesticide chemicals in or on raw agricultural commodities [346a]
Sec. 409. Food additives [348]
Sec. 410. Bottled drinking water [349]
Sec. 411. Vitamins and minerals [350]
Sec. 412. Requirements for Infant Formulas [350a]

CHAPTER V - DRUGS AND DEVICES
Subchapter A - Drugs and Devices

Sec. 501. Adulterated drugs and devices [351]
Sec. 502. Misbranded drugs and devices [352]
Sec. 503. Exemptions in case of drugs and devices [352]
Sec. 505. New drugs [355]
Sec. 506. Certification of drugs containing insulin [356]
Sec. 507. Certification of antibiotics [357]
Sec. 508. Authority to designate official names [358]
Sec. 509. Nonapplicability to cosmetics [359]
Sec. 510. Registration of producers of drugs and devices [360]
Sec. 512. New animal drugs [360b]
Sec. 513. Classification of devices intended for human use [360c]
Sec. 514. Performance standards [360d]
Sec. 515. Premarket approval [360e]
Sec. 516. Banned devices [360f]
Sec. 517. Judicial review [360g]
Sec. 518. Notification and other remedies [360h]
Sec. 519. Records and reports on devices [360i]
Sec. 520. General provisions respecting control of devices intended for human use [360j]
Sec. 521. State and local requirements respecting devices [360k]

[Sec. 522. through 524. Reserved]

Subchapter B - Drugs for Rare Diseases or Conditions

Sec. 525. Recommendations for investigations of drugs for rare diseases or conditions [360aa]
Sec. 526. Designation of drugs for rare diseases or conditions [360bb]
Sec. 527. Protection for unpatented drugs for rare diseases or conditions [360cc]
Sec. 528. Open protocols for investigations of drugs for rare diseases or conditions [360dd]

CHAPTER VI - COSMETICS

Sec. 601. Adulterated cosmetics [361]
Sec. 602. Misbranded cosmetics [362]
Sec. 603. Regulations making exemptions [363]

CHAPTER VII - GENERAL ADMINISTRATIVE PROVISIONS

Sec. 701. Regulations and hearings [371]
Sec. 702. Examinations and investigations [372]
Sec. 702a. Seafood inspection [372a]
Sec. 703. Records of interstate shipment [373]
Sec. 704. Factory inspection [374]
Sec. 705. Publicity [375]
Sec. 706. Listing and certification of color additives for foods, drugs, and cosmetics [376]
Revisions of U.S. Pharmacopeia; development of analysis and mechanical and physical tests [377]
Sec. 707. Advertising of certain foods [378]
Sec. 708. Confidential information [379]
Sec. 709. Presumption [379a]

CHAPTER VIII - IMPORTS AND EXPORTS

Sec. 801. Imports and exports [381]

CHAPTER IX - MISCELLANEOUS

SECTION I - SHORT TITLE

SEC. 1. This Act may be cited as the Federal Food, Drug, and Cosmetic Act.

SECTION II - DEFINITIONS

SEC. 201. [321] For the purposes of this Act-

(a) (1) The term "State", except as used in the last sentence of section 702(a), means any State or Territory of the United States, the District of Columbia, and the Commonwealth of Puerto Rico.

(2)[1] The term "Territory" means any Territory or possession of the United States, including the District of Columbia, and excluding the Commonwealth of Puerto Rico and the Canal Zone.

---

[1]Subsec. 201(a)(2) amended by sec. 4(a) of P.L. 90-639 and by Sec. 701 of P.L. 91-513.
*The following additional definitions for food are provided for in other acts:

SEC. 201a [321a] Butter. The Act of March 4, 1923 (42 Stat. 1500), defines butter as "For the purposes of this chapter 'butter' shall be understood to mean the food product usually known as butter, and which is made exclusively from milk or cream, or both, with or without common salt, and with or without additional coloring matter, and containing not less than 80 per centum by weight of milk fat, all tolerances having been allowed for."

SEC. 201b [321b] Package. The Act of July 24, 1919 (41 Stat. 271), declares "The word 'package' where it occurs in this chapter shall include and shall be construed to include wrapped meats enclosed in papers or other materials as prepared by the manufacturers thereof for sale."

SEC 201c [321c] Nonfat Dry Milk. The Act of July 1, 1956 (70 Stat. 486), defines nonfat dry milk as follows: *for the purposes of the Federal Food, Drug, and Cosmetic Act of June 26 sic. 1938 (ch. 675, sec. 1, 52 Stat. 1040), nonfat dry milk is the product resulting from the removal of fat and water from milk, and contains the lactose, milk proteins, and milk minerals in the same relative proportions as in the fresh milk from which made. It contains not over 5 per centum by weight of moisture. The fat content is not over 1½ per centum by weight unless otherwise indicated."

"The term 'milk', when used herein, means sweet milk of cows."

The definition of oleomargine appears preceding sec. 407a.

(b) The term "interstate commerce" means

(1) commerce between any State or Territory and any place outside thereof, and

(2) commerce within the District of Columbia or within any other Territory not organized with a legislative body.

(c) The term "Department" means the U.S. Department of Health and Human Services.

(d) The term "Secretary" means the Secretary of Health and Human Services.

(e) The term "person includes individual, partnership, corporation, and association.

(f)* The term "food" means

(1) articles used for food or drink for man or other animals,

(2) chewing gum, and

(3) articles used for components of any other such article.

(g) (1) The term "drug" means

(A) articles recognized in the official United States Pharmacopeia, official Homeopathic Pharmacopeia of the United States, or official National Formulary, or any supplement to any of them; and

(B) articles intended for use in the diagnosis, cure mitigation treatment, or prevention of disease in man or other animal, and

(C) articles (other than food) intended to affect the structure or any function of the body of man or other animals; and

(D) articles intended for use as a component of any articles specified in clause (A), (B), or (C); but does not include devices or their components, parts, or accessories.

(2) The term "counterfeit drug" means a drug which, or the container or labeling of which, without authorization, bears the trademark, trade name, or other identifying mark, imprint, or device, or any likeness thereof, of a drug manufacturer, processor, packer, or distributor other than the person or persons who in fact manufactured, processed, packed, or distributed such drug and which thereby falsely purports or is represented to be the

product of, or to have been packed or distributed by, such other drug manufacturer, processor, packer, or distributor.

(h)[2] The term "device" (except when used in paragraph (n) of this section and in sections 301 (i), 403 (f), 502 (c), and 602 (c)) means an instrument, apparatus, implement, machine, contrivance, implant, in vitro reagent, or other similar or related article, including any component, part, or accessory, which is–

(1) recognized in the official National Formulary, or the United States Pharmacopeia, or any supplement to them,

(2) intended for use in the diagnosis of disease or other conditions, or in the cure, mitigation, treatment, or prevention of disease, in man or other animals, or

(3) intended to affect the structure or any function of the body of man or other animals, and which does not achieve any of its principal intended purposes through chemical action within or on the body of man or other animals and which is not dependent upon being metabolized for the achievement of any of its principal intended purposes.

(i) The term "cosmetic" means

(1) articles intended to be rubbed, poured, sprinkled, or sprayed on, introduced into, or otherwise applied to the human body or any part thereof for cleansing, beautifying, promoting attractiveness, or altering the appearance, and

(2) articles intended for use as a component of any such articles; except that such term shall not include soap.

(j) The term "official compendium" means the official United States Pharmacopeia, official Homeopathic Pharmacopeia of the United States, official National Formulary, or any supplement to any of them.

(k) The term "label" means a display of written, printed, or graphic matter upon the immediate container of any article; and a requirement made by or under authority of this Act that any word, statement, or other information appear on the label shall not be considered to be complied with unless such word, state-

[2] Sec. 201(h) amended by sec. 3(a)(1) of P.L. 94-295.

ment, or other information also appears on the outside container or wrapper, if any there be, of the retail package of such article, or is easily legible through the outside container or wrapper.

(l) The term "immediate container" does not include package liners.

(m) The term "labeling" means all labels and other written printed, or graphic matter (1) upon any article or any of its containers or wrappers, or (2) accompanying such article.

(n)[3] If an article is alleged to be misbranded because the labeling or advertising is misleading, then in determining whether the labeling or advertising is misleading there shall be taken into account (among other things)not only representations made or suggested by statement, word, design, device, or any combination thereof, but also the extent to which the labeling or advertising fails to reveal facts material in the light of such representations or material with respect to consequences which may result from the use of the article to which the labeling or advertising relates under the conditions of use prescribed in the labeling or advertising thereof or under such conditions of use as are customary or usual.

(o) The representation of a drug, in its labeling, as an antiseptic shall be considered to be a representation that is a germicide, except in the case of a drug purporting to be, or represented as, an antiseptic for inhibitory use as a wet dressing, ointment, dusting powder, or such other use as involves prolonged contact with the body.

(p)[4] The term "new drug" means:

---

[3] Sec. 201(n) amended by sec. 501(a)(2)(A) of P.L. 94-278.

[4] Sec. 201(p) amended by secs. 102(a) and (b) of P.L. 90-399.

"ENACTMENT DATE" AS DEFINED IN DRUG AMENDMENTS ACT OF 1962
Sec. 107 of P.L. 87-781

Sec. 107(c)(1) §As used in this subsection, the term "enactment date" means the date of enactment of this Act; and the term "basic Act" means the Federal Food, Drug, and Cosmetic Act.

(2) An application filed pursuant to section 505(b) of the basic Act which was "effective" within the meaning of that Act on the date immediately preceding the enactment date shall be deemed, as of the enactment date, to be an application "approved" by the Secretary within the meaning of the basic Act as amended by this Act.

(3) In the case of any drug with respect to which an application filed under section 505(b) of the basic Act is deemed to be an approved application on the enactment date by virtue of paragraph (2) of this subsection–

CONTINUED BOTTOM NEXT PAGE

(1) Any drug (except a new animal drug or an animal feed bearing or containing a new animal drug) the composition of which is such that such drug is not generally recognized, among experts qualified by scientific training and experience to evaluate the safety and effectiveness of drugs, as safe and effective for use under the conditions prescribed, recommended, or suggested in the labeling thereof, except that such a drug not so recognized shall not be deemed to be a "new drug" if at any time prior to the enactment of this Act it was subject to the Food and Drugs Act of June 30, 1906, as amended, and if at such time its labeling contained the same representations concerning the conditions of its use; or,

(2) Any drug (except a new animal drug or an animal feed bearing or containing a new animal drug) the composition of which is such that such drug, as a result of investigations to determine its safety and effectiveness for use under such conditions, has become so recognized, but which has not, other-

---

CONTINUED FROM PRECEDING PAGE

(A) the amendments made by this Act to section 201(p), and to subsections (b) and (d) of section 505, of the basic Act, insofar as such amendments relate to the effectiveness of drugs, shall not, so long as approval of such application is not withdrawn or suspended pursuant to section 505(c) of the Act, apply to such drug when intended solely for use under conditions prescribed, recommended, or suggested in labeling covered by such approved application, but shall apply to any changed use, or conditions of use, prescribed, recommended, or suggested in its labeling, including such conditions of use as are the subject of an amendment or supplement to such application pending on, or filed after, the enactment date; and

(B) clause(3) of the first sentence of section 505(e) of the basic Act, as amended by this Act, shall not apply to such drug when intended solely for use under conditions prescribed, recommended, or suggested in labeling covered by such approved application (except with respect to such use, or conditions of use, as are the subject of an amendment or supplement to such approved application, which amendment or supplement has been approved after the enactment date under section 505 of the basic Act as amended by this Act) until whichever of the following first occurs: (i) the expiration of the two-year period beginning with the enactment date: (ii) the effective date of an order under section 505(e) of the basic Act, other than clause (3) of the first sentence of such section 505(e), withdrawing or suspending the approval of such application.

(4) In the case of any drug which, on the first day immediately preceding the enactment date, (A) was commercially used or sold in the United States, (b) was not a new drug as defined by section 201(p) of the basic Act as then in force, and (C) was not covered by an effective application under section 505 of that Act, the amendments to section 201(p) made by this Act shall not apply to such drug when intended solely for the use under conditions prescribed, recommended, or suggested in labeling with respect to such drug on the day.

wise than in such investigations, been used to a material extent or for a material time under such conditions.

(q)The term "pesticide chemical" means any substance which, alone, in chemical combination or in formulation with one or more other substance, is "a pesticide" within the meaning of the Federal Insecticide, Fungicide, and Rodenticide Act (7 U.S.C., sec. 136(u)) as now in force or as hereafter amended, and which is used in the production, storage, or transportation of raw agricultural commodities.

(r) The term "raw agricultural commodity" means any food in its raw or natural state, including all fruits that are washed, colored, or otherwise treated in their unpeeled natural form prior to marketing.

(s)[5] The term "food additive" means any substance the intended use of which results or may reasonably be expected to result, directly or indirectly, in its becoming a component or otherwise affecting the characteristics of any food (including any substance intended for use in producing, manufacturing, packing, processing, preparing, treating, packaging, transporting, or holding food; and including any source of radiation intended for any such use), if such substance is not generally recognized, among experts qualified by scientific training and experience to evaluate its safety, as having been adequately shown through scientific procedures (or, in the case of a substance used in food prior to January 1, 1958, through either scientific procedures or experience based on common use in food) to be safe under the conditions of its intended use; except that such term does not include–

(1) a pesticide chemical in or on a raw agricultural commodity; or

(2) a pesticide chemical to the extent that it is intended for use or is used in the production, storage, or transportation of any raw agricultural commodity; or

(3) a color additive; or

(4) any substance used in accordance with a sanction or approval granted prior to the enactment of this paragraph pursuant to this Act, the Poultry Products Inspection Act (21 U.S.C. 451 and the following) or the Meat inspection Act of March 4, 1907 (34 Stat. 1260), as

[5] Subsec. 201(s)(5) added by sec. 102(c) of P.L. 90-399.

amended and extended (21 U.S.C. 71 and the following); or

(5) a new animal drug.

(t) (1) The term "color additive" means a material which–

(A) is a dye, pigment, or other substance made by a process of synthesis or similar artifice, or extracted, isolated, or otherwise derived, with or without intermediate or final change of identity, from a vegetable, animal, mineral, or other source, and

(B) when added or applied to a food, drug, or cosmetic, or to the human body or any part thereof, is capable (alone or through reaction with other substance) of imparting color thereto:

except that such term does not include any material which the Secretary, by regulation, determines is used (or intended to be used) solely for a purpose or purposes other than coloring.

(2) The term "color" includes black, white, and intermediate grays.

(3) Nothing in subparagraph (1) of this paragraph shall be construed to apply to any pesticide chemical, soil or plant nutrient, or other agricultural chemical solely because of its effect in aiding, retarding, or otherwise affecting, directly or indirectly, the growth or other natural physiological processes of produce of the soil and thereby affecting its color, whether before or after harvest.

(u)[6] The term "safe," as used in paragraph(s) of this section and in sections 409, 512, and 706, has reference to the health of man or animal.

(v)[7] ***

(w)[8] The term "new animal drug" means any drug intended for use for animals other than man, including any drug intended for use in animal feed but not including such animal feed–

[6] Sec. 201(u) amended by sec. 102(d) of P.L. 90-399.
[7] Subsec. 201(v) repealed by sec. 701 of P.L. 91-513.
[8] Secs. 201(w) and (x) added by sec. 102(e) of P.L. 90-399.

(1) the composition of which is such that such drug is not generally recognized, among experts qualified by scientific training and experience to evaluate the safety and effectiveness of animal drugs, as safe and effective for use under the conditions prescribed, recommended, or suggested in the labeling thereof; except that such a drug not so recognized shall not be deemed to be a "new animal drug" if at any time prior to June 25, 1938, it was subject to the Food and Drugs Act of June 30, 1906, as amended, and if at such time its labeling contained the same representations concerning the conditions of its use; or

(2) the composition of which is such that such drug, as a result of investigations to determine its safety and effectiveness for use under such conditions, has become so recognized but which has not, otherwise than in such investigations, been used to a material extent or for a material time under such conditions; or

(3) which drug is composed wholly or partly of any kind of penicillin, streptomycin, chlortetracycline, chloramphenicol, or bacitracin, or any derivative thereof, except when there is in effect a published order of the Secretary declaring such drug not to be a new animal drug on the grounds that (A) the requirements of certification of batches of such drug, as provided for in section 512(n), is not necessary to insure that the objectives specified in paragraph (3) thereof are achieved and (B) that neither subparagraph (1) nor (2) of this paragraph (w) applies to such drug.

(x)[8] The term "animal feed," as used in paragraph (w) of this section, in section 512, and in provisions of this Act referring to such paragraph or section, means an article which is intended for use for food for animals other than man and which is intended for use as a substantial source of nutrients in the diet of the animal, and is not limited to a mixture intended to be the sole ration of the animal.

---

[8] Secs. 201(w) and (x) added by sec. 102(e) of P.L. 90-399.

(y)[9] The term "informal hearing" means a hearing which is not subject to section 554, 556, or 557 of title 5 of the United States Code and which provides for the following:

(1) The presiding officer in the hearing shall be designated by the Secretary from officers and employees of the Department of Health and Human Services who have not participated in any action of the Secretary which is the subject of the hearing and who are not directly responsible to an officer or employee of the Department who has participated in any such action.

(2) Each party to the hearing shall have the right at all times to be advised and accompanied by an attorney.

(3) Before the hearing, each party to the hearing shall be given reasonable notice of the matters to be considered at the hearing, including a comprehensive statement of the basis for the action taken or proposed by the Secretary which is the subject of the hearing and a general summary of the information which will be presented by the Secretary at the hearing in support of such action.

(4) At the hearing the parties to the hearing shall have the right to hear a full and complete statement of the action of the Secretary which is the subject of the hearing together with the information and reasons supporting such action, to conduct reasonable questioning, and to present any oral or written information relevant to such action.

(5) The presiding officer in such hearing shall prepare a written report of the hearing to which shall be attached all written material presented at the hearing. The participants in the hearing shall be given the opportunity to review and correct or supplement the presiding officer's report of the hearing.

6) The Secretary may require the hearing to be transcribed. A party to the hearing shall have the right to have the hearing transcribed at his expense. Any transcription of a hearing shall be included in the presiding officer's report of the hearing.

---

[9] Sec. 201(y) added by sec. 3(a)(2) of P.L. 94-295.

(z)[10] The term "saccharin" includes calcium saccharin, sodium saccharin, and ammonium saccharin.

(aa) The term "infant formula" means a food which purports to be or is represented for special dietary use solely as a food for infants by reason of its simulation of human milk or its suitability as a complete or partial substitute for human milk.

SECTION III - PROHIBITED ACTS AND PENALTIES

PROHIBITED ACTS

SEC. 301. [331] The following acts and the causing thereof are hereby prohibited:

(a) The introduction or delivery for introduction into interstate commerce of any food, drug, device, or cosmetic that is adulterated or misbranded.

(b) The adulteration or misbranding of any food, drug, device, or cosmetic in interstate commerce.

(c) The receipt in interstate commerce of any food, drug, device, or cosmetic that is adulterated or misbranded, and the delivery or proffered delivery thereof for pay or otherwise.

(d) The introduction or delivery for introduction into interstate commerce of any article in violation of section 404 or 505.

(e)[11] The refusal to permit access to or copying of any record as required by section 412 or 703; or the failure to establish or maintain any record, or make any report, required under section 412, 505 (i) or (j), 507 (d) or (g), 512 (j), (l) or (m), 515 (f) or 519, or the refusal to permit access to or verification or copying of any such required record.

(f) The refusal to permit entry or inspection as authorized by section 704.

(g) The manufacture within any Territory of any food, drug, device, or cosmetic that is adulterated or misbranded.

(h) The giving of a guaranty or undertaking referred to in section 303(c)(2), which guaranty or undertaking is false, except

[10] Sec. 201(z) added by sec. (4)(b)(2) by P.L. 95-203.
[11] Sec. 301(e) amended by sec. 103(1) of P.L. 90-399.

by a person who relied upon a guaranty or undertaking to the same effect signed by, and containing the name and address of, the person residing in the United States from whom he received in good faith the food, drug, device, or cosmetic; or the giving of a guaranty or undertaking referred to in section 303(c)(3), which guaranty or undertaking is false.

(i) (1) Forging, counterfeiting, simulating, or falsely representing, or without proper authority using any mark, stamp, tag, label, or other identification device authorized or required by regulations promulgated under the provisions of section 404, 506, 507, or 706.

(2) Making, selling, disposing of, or keeping in possession, control, or custody, or concealing any punch, die, plate, stone or other thing designed to print, imprint, or reproduce the trademark, trade name, or other identifying mark, imprint, or device of another or any likeness of any of the foregoing upon any drug or container or labeling thereof so as to render such drug a counterfeit drug.

(3) The doing of any act which causes a drug to be a counterfeit drug, or the sale or dispensing, or the holding for sale or dispensing, of a counterfeit drug.

(j)[12] The using by any person to his own advantage, or revealing, other than to the Secretary or officers or employees of the Department, or to the courts when relevant in any judicial proceeding under this Act, any information acquired under authority of section 404, 409, 412, 505, 506, 507, 510, 512, 513, 514, 515, 516, 518, 519, 520, 704, 706, or 708 concerning any method or process which as a trade secret is entitled to protection.

(k) The alteration, mutilation, destruction, obliteration, or removal of the whole or any part of the labeling of, or the doing of any other act with respect to, a food, drug, device, or cosmetic, if such act is done while such article is held for sale (whether or not the first sale) after shipment in interstate commerce and results in such article being adulterated or misbranded.

(l) The using, on the labeling of any drug or device or in any

[12] Sec. 301(j) amended by sec. 103(2) of P.L. 90-399.

advertising relating to such drug or device, of any representation or suggestion that approval of an application with respect to such drug or device is in effect under section 505, 515, or 520(g), as the case may be, or that such drug or device complies with the provisions of such action.

(m) The sale or offering for sale of colored oleomargarine or colored margarine, or the possession or serving of colored oleomargarine or colored margarine in violation of section 407(b) or 407(c).

(n) The using, in labeling, advertising or other sales promotion of any reference to any report or analysis furnished in compliance with section 704.

(o) In the case of a prescription drug distributed or offered for sale in interstate commerce, the failure of the manufacturer, packer, or distributor thereof to maintain for transmittal, or to transmit, to any practitioner licensed by applicable State law to administer such drug who makes written request for information as to such drug, true and correct copies of all printed matter which is required to be included in any package in which that drug is distributed or sold, or such other printed matter as is approved by the Secretary. Nothing in this paragraph shall be construed to exempt any person from any labeling requirement imposed by or under other provisions of this Act.

(p) The failure to register in accordance with section 510, the failure to provide any information required by section 510(j) or 510(k), or the failure to provide a notice required by section 510(j)(2).

(q)[13] (1) The failure or refusal to (A) comply with any requirement prescribed under section 518 or 520(g), or (B) furnish any notification or other material or information required by or under section 519 or 520(g).

(2) With respect to any device, the submission of any report that is required by or under this Act that is false or misleading in any material respect.

(r) [14]The movement of a device in violation of an order under section 304(g) or the removal or alteration of any mark or label

[13] Sec. 301(q) added by sec. 3(b)(1) of P.L. 94-295.

[14] Sec. 301(r) added by sec. 7(b) of P.L. 94-295. (note: previous sec. 301(q) repealed by sec. 701 of P.L. 91-513).

required by the order to identify the device as detained.
(s) The failure to provide the notice required by section 412(b) or 412(c), the failure to make the reports required by section 412(d)(1)(B), or the failure to meet the requirements prescribed under section 412(d)(2).

INJUNCTION PROCEEDINGS

SEC. 302. [332]
(a) The district courts of the United States and the United States courts of the Territories shall have jurisdiction, for cause shown, and subject to the provisions of section 381 (relating to notice to opposite party) of title 28, to restrain violations of section 301 of this title, except paragraphs (h), (i), and (j) of said section.
(b) In case of violation of an injunction or restraining order issued under this section, which also constitutes a violation of this Act, trial shall be by the court, or, upon demand of the accused, by a jury. Such trial shall be conducted in accordance with the practice and procedure applicable in the case of proceedings subject to the provisions of section 22 of such Act of October 15, 1914, as amended. This section, which appeared as U.S.C., title 28, sec. 387, has been repealed. It is now covered by Rule 42(b), Federal Rules of Criminal Procedure.

PENALTIES

SEC. 303. [333]
(a)[15] Any person who violates a provision of section 301 shall be imprisoned for not more than one year or fined not more than $1,000, or both.
(b)[15] Not withstanding the provisions of subsection (a) of this section, if any person commits such a violation after a conviction of him under this section has become final, or commits such a violation with the intent to defraud or mislead, such person shall be imprisoned for not more than three years or fined not more than $10,000 or both.

---

[15] Secs. 303(a) and (b) amended by sec. 701 of P.L. 91-513 effective May 1, 1971.

(c) No person shall be subject to the penalties of subsection (a) of this section,

(1) for having received in interstate commerce any article and delivered it or proffered delivery of it, if such delivery or proffer was made in good faith, unless he refuses to furnish on request of an officer or employee duly designated by the Secretary the name and address of the person from whom he purchased or received such article and copies of all documents, if any there be, pertaining to the delivery of the article to him; or

(2) for having violated section 301 (a) or (d), if he establishes a guaranty or undertaking signed by, and containing the name and address of, the person residing in the United States from whom he received in good faith the article, to the effect, in case of an alleged violation of section 301(a), that such article is not adulterated or misbranded, within the meaning of this Act, designating this Act, or to the effect, in case of an alleged violation of section 301(d), that such article is not an article which may not, under the provisions of section 404 or 505, be introduced into interstate commerce; or

(3) for having violated section 301(a), where the violation exists because the article is adulterated by reason of containing a color additive not from a batch certified in accordance with regulations promulgated by the Secretary under this Act, if such person establishes a guaranty or undertaking signed by, and containing the name and address of, the manufacturer of the color additive, to the effect that such color additive was from a batch certified in accordance with the applicable regulations promulgated by the Secretary under this Act; or

(4) for having violated section 301 (b), (c), or (k) by failure to comply with section 502(f) in respect to an article received in interstate commerce to which neither section 503(a) nor section 503(b)(1) is applicable if the delivery or proffered delivery was made in good faith and the labeling at the time thereof contained the same directions for use and warning statements as were contained in the labeling at the time of such receipt of such article; or

(5) for having violated section 301(i)(2) if such person acted in good faith and had no reason to believe that use of the punch, die, plate, stone, or other thing involved would result in a drug being a counterfeit drug, or for having violated section 301(i)(3) if the person doing the act or causing it to be done acted in good faith and had no reason to believe that the drug was a counterfeit drug.

(d)[16] No person shall be subject to the penalties of subsection (a) of this section for a violation of section 301 involving misbranded food if the violation exists solely because the food is misbranded under section 403 (a) (2) because of its advertising, and no person shall be subject to the penalties of subsection (b) of this section for such a violation unless the violation is committed with the intent to defraud or mislead.

SEIZURE

SEC. 304 334 (a)[17]

(1) Any article of food, drug, or cosmetic that is adulterated or misbranded when introduced into or while in interstate commerce or while held for sale (whether or not the first sale) after shipment in interstate commerce, or which may not, under the provisions of section 404 or 505, be introduced into interstate commerce, shall be liable to be proceeded against while in interstate commerce, or at any time thereafter, on libel of information and condemned in any district court of the United States or United States court of a Territory within the jurisdiction of which the article is found: *Provided, however,* That no libel for condemnation shall be instituted under this Act, for any alleged misbranding if there is pending in any court a label for condemnation proceeding under this Act based upon the same alleged misbranding, and not more than one such proceeding shall be instituted if no such proceeding is so pending, except that such limitations shall not apply (A) when such misbranding has

---

[16] Sec. 303(d) added by sec. 502(a)(2)(b) of P.L. 94-278.

[17] Sec. 304(a) amended by sec. 4(b) of P.L. 90-639, secs. 304(a)(2) and (d)(3) by sec. 701 of P.L. 91-513, and sec. 304(a)(1) by sec. 3(c) of P.L. 94-295.

been the basis of a prior judgment in favor of the United States, in a criminal, injunction, or libel for condemnation proceeding under this Act, or (B) when the Secretary has probable cause to believe from facts found, without hearing, by him or any officer of employee of the Department that the misbranded article is dangerous to health, or that the labeling of the misbranded article is fraudulent, or would be in a material respect misleading to the injury or damage of the purchaser or consumer. In any case where the number of libel for condemnation proceedings is limited as above provided the proceeding pending or instituted shall, on application on the claimant, reasonably made, be removed for trial to any district agreed upon by stipulation between the parties, or, in case of failure to so stipulate within a reasonable time, the claimant may apply to the court of the district in which the seizure has been made, and such court (after giving the United States attorney for such district reasonable notice and opportunity to be heard) shall by order, unless good cause to the contrary is shown, specify a district of reasonable proximity to the claimant's principal place of business to which the case shall be removed for trial.

(2) The following shall be liable to be proceeded against at any time on libel of information and condemned in any district court of the United States or United States court of a Territory within the jurisdiction of which they are found: (A) any drug that is a counterfeit drug, (B) any container of a counterfeit drug, (C) any punch, die, plate, stone, labeling, container, or other thing used or designed for use in making a counterfeit drug or drugs, and (D) any adulterated or misbranded device.

(3)[18](A) Except as provided in subparagraph (B), no libel for condemnation may be instituted under paragraph (1) or (2) against any food which (i) is misbranded under section 403(a)(2) because of its advertising, and (ii) is

[18] Sec. 304(a)(3) added by sec. 502(a)(2)(C) of P.L. 94-278.

being held for sale to the ultimate consumer in an establishment other than an establishment owned or operated by a manufacturer, packer, or distributor of the food. (B) A libel for condemnation may be instituted under paragraph (1) or (2) against a food described in subparagraph (A) if–

(i)–

(I) the food's advertising which resulted in the food being misbranded under section 403(a) (2) was disseminated in the establishment in which the food is being held for sale to the ultimate consumer,

(II) such advertising was disseminated by, or under the direction of, the owner or operator of such establishment, or

(III) all or part of the cost of such advertising was paid by such owner or operator; and

(ii) the owner or operator of such establishment used such advertising in the establishment to promote the sale of the food.

(b) The article, equipment, or other thing proceeded against shall be liable to seizure by process pursuant to the libel, and the procedure in cases under this section shall conform, as nearly as may be, to the procedure in admiralty; except that on demand of either party any issue of fact joined in any such case shall be tried by jury. When libel for condemnation proceedings under this section, involving the same claimant and the same issues of adulteration or misbranding, are pending in two or more jurisdictions, such pending proceedings, upon application of the claimant reasonably made to the court of one such jurisdiction, shall be consolidated for trial by order of such court, and tried in (1) any district selected by the claimant where one of such proceedings is pending; or (2) a district agreed upon by stipulation between the parties. If no order for consolidation is so made within a reasonable time, the claimant may apply to the court of one such jurisdiction, and such court (after giving the United States attorney for such district reasonable notice and the opportunity to be heard) shall by order, unless good cause to the contrary is shown, specify a district of reasonable proximity to the claimant's principal place of business, in which all such pending proceedings shall be consolidated for trial and tried. Such order of consolidation shall not apply so as to require the removal of any case the date for trial of which has been fixed. The court granting such order shall give prompt notification thereof to the other courts having jurisdiction of the cases covered thereby.

(c) The court at any time after seizure up to a reasonable time before trial shall by order allow any party to a condemnation proceeding, his attorney or agent, to obtain a representative sample of the article seized and a true copy of the analysis, if any, on which the proceeding is based and the identifying marks or numbers, if any, of the packages from which the samples analyzed were obtained.

(d)* –

(1) Any food, drug, device, or cosmetic condemned under this section shall, after entry of the decree, be disposed of by destruction or sale as the court may, in accordance with the provisions of this section, direct and the proceeds thereof, if sold, less the legal costs and charges, shall be paid into the Treasury of the United States; but such article shall not be sold under such decree contrary to the provisions of this Act or the laws of the jurisdiction in which sold: Provided, That after entry of the decree and upon the payment of the costs of such proceedings and the execution of a good and sufficient bond conditioned that such article shall not be sold or disposed of contrary to the provisions of this Act or the laws of any State or Territory in which sold, the court may by order direct that such article be delivered to the owner thereof to be destroyed or brought into compliance with the provisions of this Act under the supervision of an officer or employee duly designated by the Secretary, and the expenses of such supervision shall be paid by the person obtaining release of the article under bond. If the article was imported into the United States and the person seek-

*Through an apparent oversight, Congress did not amend this provision to conform to amendments made by P.L. 94-295 to the export provisions of §801(d). Reference to ". . .clauses (1) and (2) of section 801(d). . ." should read ". . .clauses (1)(A) and (B) of section 801(d). . ."

ing its release establishes (A) that the adulteration, misbranding, or violation did not occur after the article was imported, and (B) that he had no cause for believing that it was adulterated, misbranded, or in violation before it was released from customs custody, the court may permit the article to be delivered to the owner for exportation in lieu of destruction upon a showing by the owner that all of the conditions of section 801(d) can and will be met: *Provided, however,* That the provisions of this sentence shall not apply where condemnation is based upon violation of section 402(a) (1), (2), or (6), section 501 (a) (3), section 502(j), or section 601 (a) or (d); *And provided further,* That where such exportation is made to the original foreign supplier, then clauses (1) and (2) of section 801(d) and the foregoing proviso shall not be applicable; and in all cases of exportation the bond shall be conditioned that the article shall not be sold or disposed of until the applicable conditions of section 801(d) have been met. Any article condemned by reason of its being an article which may not, under section 404 or 505, be introduced into interstate commerce, shall be disposed of by destruction.

(2) The provisions of paragraph (1) of this subsection shall, to the extent deemed appropriate by the court, apply to any equipment or other thing that is not otherwise within the scope of such paragraph and which is referred to in paragraph (2) of subsection (a).

(e) Whenever in any proceeding under this section, involving paragraph (2) of subsection (a), the condemnation of any equipment or thing (other than a drug) is decreed, the court shall allow the claim of any claimant, to the extent of such claimant's interest, for remission or mitigation of such forfeiture if such claimant proves to the satisfaction of the court (i) that he has not committed or caused to be committed any prohibited act referred to in such paragraph (2) and has no interest in any drug referred to therein, (ii) that he has an interest in such equipment or other thing as owner or lienor or otherwise, acquired by him in good faith, and (iii) that he at no time had any knowledge or reason to believe that such equipment or other thing was being or would be used in, or to facilitate, the violation of laws of the United States relating to counterfeit drugs.

When a decree of condemnation is entered against the article, court costs and fees, and storage and other proper expenses, shall be awarded against the person, if any, intervening as claimant of the article.

(f) In the case of removal for trial of any case as provided by subsection (a) or (b)

(1) The clerk of the court from which removal is made shall promptly transmit to the court in which the case is to be tried all records in the case necessary in order that such court may exercise jurisdiction.

(2) The court to which such case was removed shall have the powers and be subject to the duties for purposes of such case, which the court from which removal was made would have had, or to which such court would have been subject, if such case had not been removed.

(g)[19](1) If during an inspection conducted under section 704 of a facility or a vehicle, a device which the officer or employee making the inspection has reason to believe is adulterated or misbranded is found in such facility or vehicle, such officer or employee may order the device detained (in accordance with regulations prescribed by the Secretary) for a reasonable period which may not exceed twenty days unless the Secretary determines that a period of detention greater than twenty days is required to institute an action under subsection (a) or section 302, in which case he may authorize a detention period of not to exceed thirty days. Regulations of the Secretary prescribed under this paragraph shall require that before a device may be ordered detained under this paragraph shall require the Secretary or an officer or employee designated by the Secretary approve such order. A detention order under this paragraph may require the labeling or marking of a device during the period of its detention for the purpose of identifying the device as detained. Any person who would be entitled to claim a device if it were seized under subsection (a) may appeal to the Secretary a detention of such device under this paragraph. Within five

[19] Sec. 304(g) added by sec. 7(a) of P.L. 94-295.

days of the date of an appeal of a detention is filed with the Secretary, the Secretary shall after affording opportunity for an informal hearing by order confirm the detention or revoke it.

(2) (A) Except as authorized by subparagraph (B), a device subject to a detention order issued under paragraph (1) shall not be moved by any person from the place at which it is ordered detained until

(i) released by the Secretary, or

(ii) the expiration of the detention period applicable to such order, whichever occurs first.

(B) A device subject to a detention order under paragraph (1) may be moved

(i) in accordance with regulations prescribed by the Secretary, and

(ii) if not in final form for shipment, at the discretion of the manufacturer of the device for the purpose of completing the work required to put it in such form.

HEARING BEFORE REPORT OF CRIMINAL VIOLATION

SEC. 305. [335] Before any violation of this Act is reported by the Secretary to any United States attorney for institution of a criminal proceeding, the person against whom such proceeding is contemplated shall be given appropriate notice and an opportunity to present his views, either orally or in writing, with regard to such contemplated proceeding.

REPORT OF MINOR VIOLATIONS

SEC. 306 [336] Nothing in this Act shall be construed as requiring the Secretary to report for prosecution, or for the institution of libel or injunction proceedings, minor violations of this Act whenever he believes that the public interest will be adequately served by a suitable written notice or warning.

PROCEEDINGS IN THE NAME OF UNITED STATES; PROVISION AS TO SUBPOENAS

SEC. 307. [337] All such proceedings for the enforcement, or to restrain violations, of this Act shall be by and in the name of the United States. Subpoenas for witnesses who are required to attend a court of the United States, in any district, may run into any other district in any such proceeding.

SECTION IV - FOOD

DEFINITIONS AND STANDARDS FOR FOOD

SEC. 401. [341] Whenever in the judgement of the Secretary such action will promote honesty and fair dealing in the interest of consumers, he shall promulgate regulations fixing and establishing for any food, under its common or usual name so far as practicable, a reasonable definition and standard of identity, a reasonable standard of quality, and/or reasonable standards of fill of container: *Provided,* That no definition and standard of identity and no standard of quality shall be established for fresh or dried fruits, fresh or dried vegetables, or butter, except that definitions and standards of identity may be established for avocados, cantaloupes, citrus fruits, and melons. In prescribing any standard of fill of container, the Secretary shall give due consideration to the natural shrinkage in storage and in transit of fresh natural food and the need for the necessary packing and protective material. In the prescribing of any standard of quality for any canned fruit or canned vegetable, consideration shall be given and due allowance made for the differing characteristics of the several varieties of such fruit or vegetable. In prescribing a definition and standard of identity for any food or class of food in which optional ingredients are permitted, the Secretary shall, for the purpose of promoting honesty and fair dealing in the interest of consumers, designate the optional ingredients which shall be named on the label. Any definition and standard of identity prescribed by the Secretary for avocados, cantaloupes, citrus fruits, or melons shall relate only to maturity and to the effects of freezing.

ADULTERATED FOOD

SEC. 402. [342] A food shall be deemed to be adulterated–
(a)–

(1) if it bears or contains any poisonous or deleterious substance which may render it injurious to health; but in case the substance is not an added substance such food shall not be considered adulterated under this clause if the quantity of such substance in such food does not ordinarily render it injurious to health; or

(2)[20](A) if it bears or contains any added poisonous or added deleterious substance (other than one which is (i) a pesticide chemical in or on a raw agricultural commodity, (ii) a food additive, (iii) a color additive, or (iv) a new animal drug) which is unsafe within the meaning of section 406; or (B) if it is a raw agricultural commodity and it bears or contains a pesticide chemical which is unsafe within the meaning of section 408(a); or (C) if it is, or it bears or contains, any food additive which is unsafe within the meaning of section 409: *Provided,* That where a pesticide chemical has been used in or on a raw agricultural commodity in conformity with an exemption granted or a tolerance prescribed under section 408 and such raw agricultural commodity has been subjected to processing such as canning, cooking, freezing, dehydrating, or milling, the residue of such pesticide chemical remaining in or on such processed food shall, notwithstanding the provisions of sections 406 and 409, not be deemed unsafe if such residue in or on the raw agricultural commodity has been removed to the extent possible in good manufacturing practice and the concentration of such residue in the processed food when ready to eat is not greater than the tolerance prescribed for the raw agricultural commodity; or (D) if it is, or it bears or contains, a new animal drug (or conversion product thereof) which is unsafe within the meaning of section 512; or

---

[20] Sec. 402(a) amended by sec. 104 of P.L. 90-399.

(3) if it consists in whole or in part of any filthy, putrid, or decomposed substance, or it is otherwise unfit for food; or

(4) if it has been prepared, packed, or held under insanitary conditions whereby it may have become contaminated with filth, or whereby it may have been rendered injurious to health; or

(5) if it is, in whole or in part, the product of a diseased animal or of an animal which has died otherwise than by slaughter; or

(6) if its container is composed, in whole or in part, of any poisonous or deleterious substance which may render the contents injurious to health; or

(7) if it has been intentionally subjected to radiation, unless the use of the radiation was in conformity with a regulation or exemption in effect pursuant to section 409.

(b)–

(1) If any valuable constituent has been in whole or in part omitted or abstracted therefrom; or (2) if any substance has been substituted wholly or in part therefor; or

(3) if damage or inferiority has been concealed in any manner; or

(4) if any substance has been added thereto or mixed or packed therewith so as to increase its bulk or weight, or reduce its quality or strength, or make it appear better or of greater value than it is.

(c) If it is, or it bears or contains, a color additive which is unsafe within the meaning of section 706(a).

(d)[21] If it is confectionery, and

(1) has partially or completely imbedded therein any non-nutritive object; *Provided,* That this clause shall not apply in the case of any non-nutritive object if, in the judgement of the Secretary as provided by regulations, such object is of practical functional value to the confectionery product and would not render the product injurious or hazardous to health;

[21] Sec. 402(d) amended by P.L. 89-477.

(2) bears or contains any alcohol other than alcohol not in excess of one-half or 1% by volume derived solely from the use of flavoring extracts; or

(3) bears or contains any nonnutritive substance: *Provided,* That this clause shall not apply to a nonnutritive substance which is in or on confectionery by reason of its use for some practical functional purpose in the manufacture, packaging, or storage of such confectionery if the use of the substance does not promote deception of the consumer or otherwise result in adulteration or misbranding in violation of any provision of this Act: *And provided further,* That the Secretary may, for the purpose of avoiding or resolving uncertainty as to the application of this clause, issue regulations allowing or prohibiting the use of particular nonnutritive substances.

(e) If it is oleomargarine or margarine or butter and any of the raw material used therein consisted in whole or in part of any filthy, putrid, or decomposed substance, or such oleomargarine or margarine or butter is otherwise unfit for food.

MISBRANDED FOOD

SEC. 403. [343] A food shall be deemed to be misbranded–

(a)[22] If

(1) its labeling is false or misleading in any particular, or

(2) in the case of a food to which section 411 applies, its advertising is false or misleading in a material respect or its labeling is in violation of sec. 411(b)(2).

(b) It is offered for sale under the name of another food.

(c) If it is an imitation of another food, unless its label bears, in type of uniform size and prominence, the word "imitation" and, immediately thereafter, the name of the food imitated.

(d) If its container is so made, formed, or filled as to be misleading.

(e) If in package form unless it bears a label containing

(1) the name and place of business of the manufacturer, packer, or distributor; and

---

[22] Sec. 403(a)(2) added by sec. 502(a)(1) P.L. 94-278.

(2) an accurate statement of the quantity of the contents in terms of weight, measure, or numerical count: *Provided,* That under clause (2) of this paragraph reasonable variations shall be permitted, and exemptions as to small packages shall be established, by regulations prescribed by the Secretary.

(f) If any word, statement, or other information required by or under authority of this Act to appear on the label or labeling is not prominently placed thereon with such conspicuousness (as compared with other words, statements, designs, or devices, in the labeling) and in such terms as to render it likely to be read and understood by the ordinary individual under customary conditions of purchase and use.

(g) If it purports to be or is represented as a food for which a definition and standard of identity has been prescribed by regulations as provided by section 401, unless (1) it conforms to such definition and standard, and (2) its label bears the name of the food specified in the definition and standard, and, insofar as may be required by such regulations, the common names of optional ingredients (other than spices, flavoring, and coloring) present in such food.

(h) If it purports to be or is represented as

(1) a food for which a standard of quality has been prescribed by regulations as provided by section 401, and its quality falls below such standard, unless its label bears, in such manner and form as such regulations specify, a statement that it falls below such standard; or

(2) a food for which a standard or standards of fill of container have been prescribed by regulations as provided by section 401, and it falls below the standard of fill of container applicable thereto, unless its label bears, in such manner and form as such regulations specify, a statement that it falls below such standard.

(i) If it is not subject to the provisions of paragraph (g) of this section unless its label bears (1) the common or usual name of the food, if any there be, and (2) in case it is fabricated from two or more ingredients, the common or usual name of each such ingredient; except that spices, flavorings, and colorings, other than those sold as such, may be designated as spices, flavorings, and colorings without naming each; *Provided,* That, to the extent that compliance with the requirements of clause (2) of this paragraph is impracticable, or results in deception or unfair competition, exemptions shall be established by regulations promulgated by the Secretary.

(j) If it purports to be or is represented for special dietary uses, unless its label bears such information concerning its vitamin, mineral, and other dietary properties as the Secretary determines to be, and by regulations prescribes, as necessary in order fully to inform purchasers as to its value for such uses.

(k) If it bears or contains any artificial flavoring, artificial coloring, or chemical preservative, unless it bears labeling stating that fact: *Provided,* That to the extent that compliance with the requirements of this paragraph is impracticable, exemptions shall be established by regulations promulgated by the Secretary. The provisions of this paragraph and paragraphs (g) and (i) with respect to artificial coloring shall not apply in the case of butter, cheese, or ice cream. The provisions of this paragraph with respect to chemical preservatives shall not apply to a pesticide chemical when used in or on a raw agricultural commodity which is the produce of the soil.

(l) If it is a raw agricultural commodity which is the produce of the soil, bearing or containing a pesticide chemical applied after harvest, unless the shipping container of such commodity bears labeling which declares the presence of such chemical in or on such commodity and the common or usual name and the function of such chemical: *Provided, however,* That no such declaration shall be required while such commodity, having been removed from the shipping container, is being held or displayed for sale at retail out of such container in accordance with the custom of the trade.

(m) If it is a color additive, unless its packaging and labeling are in conformity with such packaging and labeling requirements, applicable to such color additive, as may be contained in regulations issued under section 706.

(n)[23] If its packaging or labeling is in violation of an applicable regulation issued pursuant to section 3 or 4 of the Poison Prevention Packaging Act of 1970.

[23] Sec. 403(n) added by sec. 7 of P.L. 91-601.

(o)[24] (1) If it contains saccharin, unless, except as provided in subparagraph (2), its label and labeling bear the following statement: 'USE OF THIS PRODUCT MAY BE HAZARDOUS TO YOUR HEALTH. THIS PRODUCT CONTAINS SACCHARIN WHICH HAS BEEN DETERMINED TO CAUSE CANCER IN LABORATORY ANIMALS'. Such statement shall be located in a conspicuous place on such label and labeling as proximate as possible to the name of such food and shall appear in conspicuous and legible type in contrast by typography, layout, and color with other printed matter on such label and labeling.

(2) The Secretary may by regulation review and revise or remove the requirement of subparagraph (1) if the Secretary determines such action is necessary to reflect the current state of knowledge concerning saccharin.

(p)[25]–

(1) If it contains saccharin and is offered for sale, but not for immediate consumption, at a retail establishment, unless such retail establishment displays prominently, where such food is held for sale, notice (provided by the manufacturer of such food pursuant to subparagraph (2)) for consumers respecting the information required by paragraph (o) to be on food labels and labeling.

(2) Each manufacturer of food which contains saccharin and which is offered for sale by retail establishments but not for immediate consumption shall, in accordance with regulations promulgated by the Secretary pursuant to subparagraph (4), take such action as may be necessary to provide such retail establishments with the notice required by subparagraph (1).

---

[24] Sec. 403(o) added by sec. 4(a)(1) of P.L. 95-203 (enacted Nov. 23, 1977).

The amendment made by paragraph (1) shall apply only with respect to food introduced or delivered for introduction in interstate commerce on and after the 90th day after the date of enactment of this Act.

The Secretary shall report to the Committee on Human Resources of the Senate and the Committee on Interstate and Foreign Commerce of the House of Representatives any action taken under section 403(o)(2) of the Federal Food, Drug, and Cosmetic Act.

[25] Sec. 403(p) added by sec. 4(b)(1) of P.L. 95-203.

The amendment made by paragraph (1) shall apply with respect to food which is sold in retail establishments on or after the 90th day after the effective date of the regulations of the Secretary of Health and Human Services under paragraph (p)(4) of the Federal Food, Drug, and Cosmetic Act.

(3) The Secretary may by regulation review and revise or remove the requirement of subparagraph (1) if he determines such action is necessary to reflect the current state of knowledge concerning saccharin.

(4) The Secretary shall by regulation prescribe the form, text, and manner of display of the notice required by subparagraph (1) and such other matters as may be required for the implementation of the requirements of that subparagraph and subparagraph (2). Regulations of the Secretary under this subparagraph shall be promulgated after an oral hearing but without regard to the National Environmental Policy Act of 1969 and chapter 5 of title 5, United States Code. In any action brought for judicial review of any such regulation, the reviewing court may not postpone the effective date of such regulation.

EMERGENCY PERMIT CONTROL

SEC. 404. [344]

(a) Whenever the Secretary finds after investigation that the distribution in interstate commerce of any class of food may, by reason of contamination with micro-organisms during the manufacture, processing, or packing thereof in any locality, be injurious to health, and that such injurious nature cannot be adequately determined after such articles have entered interstate commerce, he then, and in such case only, shall promulgate regulations providing for the issuance, to manufacturers, processors, or packers of such class of food in such locality, of permits to which shall be attached such conditions governing the manufacture, processing, or packing of such class of food, for such temporary period of time, as may be necessary to protect the public health; and after the effective date of such regulations, and during such temporary period, no person shall introduce or deliver for introduction into interstate commerce any such food manufactured, processed, or packed by any such manufacturer, processor, or packer unless such manufacturer, processor, or packer holds a permit issued by the Secretary as provided by such regulations.

(b) The Secretary is authorized to suspend immediately upon notice any permit issued under authority of this section if it is found that any of the conditions of the permit have been violated. The holder of a permit so suspended shall be privileged at any time to apply for the reinstatement of such permit, and the Secretary shall, immediately after prompt hearing and an inspection of the establishment, reinstate such permit if it is found that adequate measures have been taken to comply with and maintain the conditions of the permit, as originally issued or as amended.

(c) Any officer or employee duly designated by the Secretary shall have access to any factory or establishment, the operator of which holds a permit from the Secretary, for the purpose of ascertaining whether or not the conditions of the permit are being complied with, and denial of access for such inspection shall be ground for suspension of the permit until such access is freely given by the operator.

REGULATIONS MAKING EXEMPTIONS

SEC. 405. [345] The Secretary shall promulgate regulations exempting from any labeling requirement of this Act (1) small open containers of fresh fruits and fresh vegetables and (2) food which is in accordance with the practice of the trade, to be processed, labeled, or repacked in substantial quantities at establishments other than those where originally processed or packed, on condition that such food is not adulterated or misbranded under the provisions of this Act upon removal from such processing, labeling, or repacking establishment.

TOLERANCES FOR POISONOUS INGREDIENTS IN FOOD

SEC. 406. [346] Any poisonous or deleterious substance added to any food, except where such substance is required in the production thereof or cannot be avoided by good manufacturing practice shall be deemed to be unsafe for purposes of the application of clause (2)(A) of section 402(a); but when such substance is so required or cannot be so avoided, the Secretary shall promulgate regulations limiting the quantity therein or thereon to such extent as he finds necessary for the protection of public health, and any quantity exceeding the limits so fixed shall also be deemed to be unsafe for purposes of the application of clause (2)(A) of section 402(a). While such a regulation is in effect limiting the quantity of any such substance in the case of any food, such food shall not, by reason of bearing or containing any added amount of such substance, be considered to be adulterated within the meaning of clause (1) of section 402(a). In determining the quantity of such added substance to be tolerated in or on different articles of food the Secretary shall take into account the extent to which the use of such substance is required or cannot be avoided in the production of each such article, and the other ways in which the consumer may be affected by the same or other poisonous or deleterious substances.

OLEOMARGARINE OR MARGARINE

[Public Law 459 - 81st Congress,- March 16, 1950 (64 Stat. 20), amended section 15 of the Federal Trade Commission Act, As Amended, by adding at the end thereof the following new subsection: "(f) For the purposes of this section and section 407 of the Federal Food, Drug, and Cosmetic Act, As Amended, the term 'oleomargarine' or 'margarine' includes–

(1) all substances, mixtures, and compounds known as oleomargarine or margarine;

(2) all substances, mixtures, and compounds which have a consistence similar to that of butter and which contain any edible oils or fats other than milk fat if made in imitation or semblance of butter."]

[In repealing section 2301 of the Internal Revenue Code (relating to the tax on oleomargarine) Public Law 459 declared, in part: "The Congress hereby finds and declares that the sale, or the serving in public eating places, of colored oleomargarine or colored margarine without clear identification as such or which is otherwise adulterated or misbranded within the meaning of the Federal Food, Drug and Cosmetic Act depresses the market in interstate commerce for butter and for oleomargarine or margarine clearly identified and neither adulterated nor misbranded, and constitutes a burden on interstate commerce in such articles. Such burden exists, irrespective of whether such oleomargarine or margarine originates from an interstate source or from the State in which it is sold."]

["Nothing in this Act shall be construed as authorizing the possession, sale, or serving of colored oleomargarine or colored margarine in any State or Territory in contravention of the laws of such State or Territory."]

SEC. 407. [347]

(a) Colored oleomargarine or colored margarine which is sold in the same State or Territory in which it is produced shall be subject in the same manner and to the same extent to the provisions of this Act as if it had been introduced in interstate commerce.

(b) No person shall sell, or offer for sale, colored oleomargarine or colored margarine unless

(1) such oleomargarine or margarine is packaged,

(2) the net weight of the contents of any package sold in a retail establishment is one pound or less,

(3) there appears on the label of the package (A) the word "oleomargarine" or "margarine" in type or lettering at least as large as any other type or lettering on such label, and (B) a full and accurate statement of all the ingredients contained in such oleomargarine, or margarine, and

(4) each part of the contents of the package is contained in a wrapper which bears the word "oleomargarine" or "margarine" in type or lettering not smaller than 20-point type.

The requirements of this subsection shall be in addition to and not in lieu of any of the other requirements of this Act.

(c) No person shall possess in a form ready for serving colored oleomargarine or colored margarine at a public eating place unless a notice that oleomargarine or margarine is served is displayed prominently and conspicuously in such place and in such manner as to render it likely to be read and understood by the ordinary individual being served in such eating place or is printed or is otherwise set forth on the menu in type or lettering not smaller than that normally used to designate the serving of other food items. No person shall serve colored oleomargarine or colored margarine at a public eating place, whether or not any charge is made therefor, unless (1) each separate serving bears or is accompanied by labeling identifying it as oleomargarine or margarine, or (2) each separate serving thereof is triangular in shape.

(d) Colored oleomargarine or colored margarine when served with meals at a public eating place shall at the time of such service be exempt from labeling requirements of section 402 of this title (except (a) and (f) if it complies with the requirements of subsection (b) of this section.

(e) For the purpose of this section colored oleomargarine or colored margarine is oleomargarine or margarine having a tint or shade containing more than one and six-tenths degrees of yellow, or of yellow and red collectively, but with an excess of yellow over red, measured in terms of Lovibond tintometer scale or its equivalent.

TOLERANCES FOR PESTICIDE CHEMICALS IN OR ON RAW AGRICULTURAL COMMODITIES

SEC. 408. [346a]

(a) Any poisonous or deleterious pesticide chemical, or any pesticide chemical which is not generally recognized, among experts qualified by scientific training and experience to evaluate the safety of pesticide chemicals, as safe for use, added to a raw agricultural commodity, shall be deemed unsafe for the purposes of the application of clause (2) of section 402(a) unless

(1) a tolerance for such pesticide chemical in or on the raw agricultural commodity has been prescribed by the Administrator of the Environmental Protection Agency under this section and the quantity of such pesticide chemical in or on the raw agricultural commodity is within the limits of the tolerance so prescribed; or

(2) with respect to use in or on such raw agricultural commodity, the pesticide chemical has been exempted from the requirement of a tolerance by the Administrator under this section.

While a tolerance or exemption from tolerance is in effect for a pesticide chemical with respect to any raw agricultural commodity, such raw agricultural commodity shall not, by reason of bearing or containing any added amount of such pesticide chemical, be considered to be adulterated within the meaning of clause (1) of section 402(a).

(b) The Administrator shall promulgate regulations establishing tolerances with respect to the use in or on raw agricultural commodities of poisonous or deleterious pesticide chemicals and of pesticide chemicals which are not generally recognized, among experts qualified by scientific training and experience to evaluate the safety of pesticide chemicals, as safe for use, to the extent necessary to protect the public health. In establishing any such regulation, the Administrator shall give appropriate consideration, among other relevant factors–

(1) to the necessity for the production of an adequate, wholesome, and economical food supply;

(2) to the other ways in which the consumer may be affected by the same pesticide chemical or by other related substances that are poisonous or deleterious, and

(3) to the opinion submitted with a certification of usefulness under subsection (1) of this section. Such regulations shall be promulgated in the manner prescribed in subsection (d) and (e) of this section. In carrying out the provisions of this section relating to the establishment of tolerances, the Administrator may establish the tolerance applicable with respect to the use of any pesticide chemical in or on any raw agricultural commodity at zero level if the scientific data before the Administrator does not justify the establishment of a greater tolerance.

(c) The Administrator shall promulgate regulations exempting any pesticide chemical from the necessity of a tolerance with respect to use in or on any or all raw agricultural commodities when such a tolerance is not necessary to protect the public health. Such regulations shall be promulgated in the manner prescribed in subsection (d) or (e) of this section.

(d)

(1) Any person who has registered, or who has submitted an application for the registration of, an economic poison

under the Federal Insecticide, Fungicide, and Rodenticide Act may file with the Administrator a petition proposing the issuance of a regulation establishing a tolerance for a pesticide chemical which constitutes, or is an ingredient of such economic poison, or exempting the pesticide chemical from the requirement of a tolerance. The petition shall contain data showing–

(A) the name, chemical identity, and composition of the pesticide chemical;

(B) the amount, frequency, and time of application of the pesticide chemical;

(C) full reports of investigations made with respect to the safety of the pesticide chemical;

(D) the results of tests on the amount of residue remaining, including a description of the analytical methods used;

(E) practicable methods for removing residue which exceeds any proposed tolerance;

(F) proposed tolerances for the pesticide chemical if tolerances are proposed; and

(G) reasonable grounds in support of the petition.

Samples of the pesticide chemical shall be furnished to the Administrator upon request. Notice of the filing of such petition shall be published in general terms by the Administrator within thirty days after filing. Such notice shall include the analytical methods available for the determination of the residue of the pesticide chemical for which tolerance or exemption is proposed.

(2) Within ninety days after a certification of usefulness by the Administrator under subsection (1) with respect to the pesticide chemical named in the petition, the Administrator shall, after giving due consideration to the data submitted in the petition or otherwise before him, by order make public a regulation–

(A) establishing a tolerance for the pesticide chemical named in the petition for the purposes for which it is so certified as useful, or

(B) exempting the pesticide chemical from the necessity of a tolerance for such purposes,

unless within such ninety-day period the person

filing the petition requests that the petition be referred to an advisory committee or the Administrator within such period otherwise deems such referral necessary, in either of which events the provisions of paragraph (3) of this subsection shall apply in lieu hereof.

(3) In the event that the person filing the petition requests, within ninety days after a certification of usefulness by the Administrator under subsection (1), with respect to the pesticide chemical named in the petition, that the petition be referred to an advisory committee, or in the event the Administrator within such period otherwise deems such referral necessary, the Administrator shall forthwith submit the petition and other data before him to an advisory committee to be appointed in accordance with subsection (g) of this section. As soon as practicable after such referral, but not later than sixty days thereafter, unless extended as hereinafter provided, the committee shall, after independent study of the data submitted to it by the Administrator and other data before it, certify to the Administrator a report and recommendations on the proposal in the petition to the Administrator, together with all underlying data and a statement of the reasons or basis for the recommendations. The sixty-day period provided for herein may be extended by the advisory committee for an additional thirty days if the advisory committee deems this necessary. Within thirty days after such certification, the Administrator shall, after giving due consideration to all data then before him, including such report, recommendations, underlying data, and statement, by order make public a regulation–

(A) establishing a tolerance for the pesticide chemical named in the petition for the purposes for which it is so certified as useful; or

(B) exempting the pesticide chemical from the necessity of a tolerance for such purposes.

(4) The regulations published under paragraph (2) or (3) of this subsection will be effective upon publication.

(5) Within thirty days after publication, any person adversely affected by a regulation published pursuant to paragraph (2) or (3) of this subsection, or pursuant to subsection (e), may file objections thereto with the Administrator, specifying with particularity the provisions of the regulation deemed objectionable, stating reasonable grounds therefor, and requesting a public hearing upon such objections. A copy of the objections filed by a person other than the petitioner shall be served on the petitioner, if the regulation was issued pursuant to a petition. The petitioner shall have two weeks to make a written reply to the objections. The Administrator shall thereupon, after due notice, hold such public hearing for the purpose of receiving evidence relevant and material to the issues raised by such objections. Any report, recommendations, underlying data, and reasons certified to the Administrator by an advisory committee shall be made a part of the record of the hearing, if relevant and material, subject to the provisions of section 7(c) of the Administrative Procedure Act (5 U.S.C., sec. 1006(c)). The National Academy of Sciences shall designate a member of the advisory committee to appear and testify at any such hearing with respect to the report and recommendations of such committee upon request of the Administrator, the petitioner, or the officer conducting the hearing: *Provided,* That this shall not preclude any other member of the advisory committee from appearing and testifying at such hearing. As soon as practicable after completion of the hearing, the Administrator shall act upon such objections and by order make public a regulation. Such regulation shall be based only on substantial evidence of record at such hearing, including any report, recommendations, underlying data, and reasons certified to the Administrator by an advisory committee, and shall set forth detailed findings of fact upon which the regulation is based. No such order shall take effect prior to the ninetieth day after its publication, unless the Administrator finds that emergency conditions exist necessitating an earlier effective date, in which

event the Administrator shall specify in the order of his findings as to such conditions.

(e) The Administrator may at any time, upon his own initiative or upon the request of any interested person, propose the issuance of a regulation establishing a tolerance for a pesticide chemical or exempting it from the necessity of a tolerance. Thirty days after publication of such a proposal, the Administrator may by order publish a regulation based upon the proposal which shall become effective upon publication unless within such thirty-day period a person who has registered, or who has submitted an application for the registration of, an economic poison under the Federal Insecticide, Fungicide, and Rodenticide Act containing the pesticide chemical named in the proposal, requests that the proposal be referred to an advisory committee. In the event of such a request, the Administrator shall forthwith submit the proposal and other relevant data before him to an advisory committee to be appointed in accordance with subsection (g) of this section. As soon as practicable after such referral, but not later than sixty days thereafter, unless extended as hereinafter provided, the committee shall, after independent study of the data submitted to it by the Administrator and other data before it, certify to the Administrator a report and recommendations on the proposal together with all underlying data and a statement of the reasons or basis for the recommendations. The sixty-day period provided for herein may be extended by the advisory committee for an additional thirty days if the advisory committee deems this necessary. Within thirty days after such certification the Administrator may, after giving due consideration to all data before him, including such report, recommendations, underlying data and statement, by order publish a regulation establishing a tolerance for the pesticide chemical named in the proposal or exempting it from the necessity of a tolerance which shall become effective upon publication. Regulations issued under this subsection shall upon publication be subject to paragraph (5) of subsection (d).

(f) All data submitted to the Administrator or to an advisory committee in support of a petition under this section shall be considered confidential by the Administrator and by such advisory committee until publication of a regulation under paragraph (2) or (3) of subsection (d) of this section. Until such publication such data shall not be revealed to any person other than those authorized by the Administrator or by an advisory committee in the carrying out of their official duties under this section.

(g) Whenever the referral of a petition or proposal to an advisory committee is requested under this section, or the Administrator otherwise deems such referral necessary, the Administrator shall forthwith appoint a committee of competent experts to review the petition or proposal and to make a report and recommendations thereon. Each such advisory committee shall be composed of experts, qualified in the subject matter of the petition and of adequately diversified professional background selected by the National Academy of Sciences and shall include one or more representatives from land-grant colleges. The size of the committee shall be determined by the Administrator. Members of an advisory committee shall receive compensation and travel expenses in accordance with subsection (b) (5) (D) of section 706.[26] [which the Administrator shall by rules and regulation prescribe.][27] The members shall not be subject to any other provisions of law regarding the appointment and compensation of employees of the United States. The Administrator shall furnish the committee with adequate clerical and other assistance, and shall by rules and regulations prescribe the procedures to be followed by the committee.

(h) A person who has filed a petition or who has requested the referral of a proposal to an advisory committee in accordance with the provision of this section, as well as representatives of the Environmental Protection Agency, shall have the right to consult with any advisory committee provided for in subsection (g) in connection with the petition or proposal.

(i)–

(1) In a case of actual controversy as to the validity of any order under subsection (d)(5), (e) or (1) any person who will be adversely affected by such order may obtain judicial review by filing in the United States Court of

[26] Subsec. 408(g) amended by sec. 601(d) of P.L. 91-515.

[27] Phrase not deleted with surrounding phrases deleted by sec. 601(d) of P.L. 91-515.

Appeals for the circuit wherein such person resides or has his principal place of business, or in the United States Court of Appeals for the District of Columbia Circuit, within sixty days after entry of such order, a petition praying that the order be set aside in whole or in part.

(2) In the case of a petition with respect to an order under subsection (d)(5) or (e), a copy of the petition shall be forthwith transmitted by the clerk of the court to the Administrator, or any officer designated by him for that purpose, and thereupon the Administrator shall file in the court the record of the proceedings on which he based his order, as provided in section 2112 of title 28, United States Code. Upon the filing of such petition, the court shall have exclusive jurisdiction to affirm or set aside the order complained of in whole or in part. The findings of the Administrator with respect to questions of fact shall be sustained if supported by substantial evidence when considered on the record as a whole, including any report and recommendation of an advisory committee.

(3) In the case of a petition with respect to an order under subsection (1), a copy of the petition shall be forthwith transmitted by the clerk of the court to the Administrator or any officer designated by him for that purpose, and thereupon the Administrator shall file in the court the record of the proceedings on which he based his order, as provided in section 2112 of title 28, United States Code. Upon the filing of such petition, the court shall have exclusive jurisdiction to affirm or set aside the order complained of in whole or in part. The findings of the Administrator with respect to questions of fact shall be sustained if supported by substantial evidence when considered on the record as a whole.

(4) If application is made to the court for leave to adduce additional evidence, the court may order such additional evidence to be taken before the Administrator, and to be adduced upon the hearing in such manner and upon such terms and conditions as the court may seem proper, if such evidence is material and there were reasonable grounds for failure to adduce such evidence in

the proceedings below. The Administrator may modify his findings as to the facts and order by reason of the additional evidence so taken, and shall file with the court such modified findings and order.

(5) the judgment of the court affirming or setting aside, in whole or in part, any order under this section shall be final, subject to review by the Supreme Court of the United States upon certiorari or certification as provided in section 1254 of title 28 of the United States Code. The commencement of proceedings under this section shall not, unless specifically ordered by the court to the contrary, operate as a stay of an order. The court shall advance on the docket and expedite the disposition of all causes filed therein pursuant to this section.

(j) The Administrator may, upon the request of any person who has obtained an experimental permit for a pesticide chemical under the Federal Insecticide, Fungicide, and Rodenticide Act or upon his own initiative, establish a temporary tolerance for the pesticide chemical for the uses covered by the permit whenever in his judgement such action is deemed necessary to protect the public health, or may temporarily exempt such pesticide chemical from a tolerance. In establishing such a tolerance, the Administrator shall give due regard to the necessity for experimental work in developing an adequate, wholesome, and economical food supply and to the limited hazard to the public health involved in such work when conducted in accordance with applicable regulations under the Federal Insecticide, Fungicide, and Rodenticide Act.

(k) Regulations affecting pesticide chemicals in or on raw agricultural commodities which are promulgated under the authority of section 406(a) upon the basis of public hearings instituted before January 1, 1953, in accordance with section 701(e), shall be deemed to be regulations under this section and shall be subject to amendment or repeal as provided in subsection (m).

(l) The Administrator, upon request of any person who has registered, or who has submitted an application for the registration of, an economic poison under the Federal Insecticide, Fungicide, and Rodenticide Act, and whose request is accompanied by a copy of a petition filed by such person under subsection (d)(1) with respect to a pesticide chemical which constitutes, or is an ingredient of, such economic poison, shall, within thirty days or within sixty days if upon notice prior to the termination of such thirty days the Administrator deems it necessary to postpone action for such period, on the basis of data before him either–

(1) certify that such pesticide chemical is useful for the purpose for which a tolerance or exemption is sought; or

(2) notify the person requesting the certification of his proposal to certify that the pesticide chemical does not appear to be useful for the purpose for which a tolerance or exemption is sought; or appears to be useful for only some of the purposes for which a tolerance or exemption is sought.

In the event that the Administrator takes the action described in clause (2) of the preceding sentence, the person requesting the certification, within one week after receiving the proposed certification, may either (A) request the Administrator to certify on the basis of the proposed certification; (B) request a hearing on the proposed certification or the parts thereof objected to; or (C) request both such certification and such hearing. If no such action is taken, the Administrator may by order make the certification as proposed. In the event that the action described in clause (A) or (C) is taken, the Administrator shall by order make the certification as proposed with respect to such parts thereof as are requested. In the event a hearing is requested, the Administrator shall provide opportunity for a prompt hearing. The certification of the Administrator as the result of such hearing shall be made by order and shall be based only on substantial evidence of record at the hearing and shall set forth detailed findings of fact. In no event shall the time elapsing between the making of a request for a certification under this subsection and final certification by the Administrator exceed one hundred and sixty days. The Administrator shall submit with any certification of usefulness under this subsection an opinion, based on the data before him whether the tolerance or exemption proposed by the petitioner reasonably reflects the amount of residue likely to result when the pesticide chemical is used in the manner proposed for the purpose for which the certification is made. The Administrator, after due notice and opportunity for public hearing, is authorized to promulgate rules and regulations for carrying out the provisions of this subsection.

(m) The Administrator shall prescribe by regulations the procedure by which regulations under this section may be amended or repealed, and such procedure shall conform to the procedure provided in this section for the promulgation of regulations establishing tolerances, including the appointment of advisory committees and the procedure for referring petitions to such committees.

(n) The provisions of section 303(c) of the Federal Food, Drug, and Cosmetic Act with respect to the furnishing of guaranties shall be applicable to raw agricultural commodities covered by this section.

(o) The Administrator shall by regulation require the payment of such fees as will in the aggregate, in the judgement of the Administrator, be sufficient over a reasonable term to provide, equip, and maintain an adequate service for the performance of the Administrator's functions under this section. Under such regulations, the performance of the Administrator's Services or other functions pursuant to this section, including any one or more of the following, may be conditioned upon the payment of such fees:

(1) the acceptance of filing of a petition submitted under subsection (d);

(2) the promulgation of a regulation establishing a tolerance, or an exemption from the necessity of a tolerance, under this section, or the amendment or repeal of such a regulation;

(3) the referral of a petition or proposal under this section to an advisory committee;

(4) the acceptance for filing of objections under subsection (d)(5); or

(5) the certification and filing in court of a transcript of the proceedings and the record under subsection (i)(2). Such regulations may further provide for waiver or refund of such fees in whole or in part when in the judge-

ment of the Administrator such waiver or refund is equitable and not contrary to the purposes of this subsection.

FOOD ADDITIVES

Unsafe Food Additives

SEC. 409. [348]

(a) A food additive shall, with respect to any particular use or intended use of such additives, be deemed to be unsafe for the purposes of the application of clause (2)(C) of section 402(a), unless

(1) it and its use or intended use conform to the terms of an exemption which is in effect pursuant to subsection (i) of this section; or

(2) there is in effect, and it and its use or intended use are in conformity with, a regulation issued under this section prescribing the conditions under which such additive may be safely used.

While such a regulation relating to a food additive is in effect, a food shall not, by reason of bearing or containing such an additive in accordance with the regulation, be considered adulterated within the meaning of clause (1) of section 402(a).

Petition to Establish Safety

(b)(1) Any person may, with respect to any intended use of a food additive, file with the Secretary a petition proposing the issuance of a regulation prescribing the conditions under which such additive may be safely used.

(2) Such petition shall, in addition to any explanatory or supporting data, contain

(A) the name and all pertinent information concerning such food additive, including, where available, its chemical identity and composition;

(B) a statement of the conditions of the proposed use of such additive, including all directions, recommendations, and suggestions proposed for the use of such additive, and including specimens of its proposed labeling;

(C) all relevant data bearing on the physical or other technical effect such additive is intended to produce, and the quantity of such additive required to produce such effect;

(D) a description of practicable methods for determining the quantity of such additive in or on food, and any substance formed in or on food, because of its use; and

(E) full reports of investigations made with respect to the safety for use of such additive, including full information as to the methods and controls used in conducting such investigations.

(3) Upon request of the Secretary, the petitioner shall furnish (or, if the petitioner is not the manufacturer of such additive, the petitioner shall have the manufacturer of such additive furnish, without disclosure to the petitioner), a full description of the methods used in, and the facilities and controls used for, the production of such additive.

(4) Upon request of the Secretary, the petitioner shall furnish samples of the food additive involved, or articles used as components thereof, and of the food in or on which the additive is proposed to be used.

(5) Notice of the regulation proposed by the petitioner shall be published in general terms by the Secretary within thirty days after filing.

Action on the Petition

(c)(1) the Secretary shall

(A) by order establish a regulation (whether or not in accord with that proposed by the petitioner) prescribing, with respect to one of more proposed uses of the food additive involved, the conditions under which such additive may be safely used (including, but not limited to, specifications as to the particular food or classes of food in or on which such additive may be used, the maximum quantity which may be used or permitted to remain in or on such food, the manner in which such additive may be added to or used in or on such food, and any directions or other labeling or packaging requirements for such additive deemed necessary by him to assure the safety of such use), and shall notify the petitioner of such order and the reasons for such action; or

(B) by order deny the petition, and shall notify the petitioner of such order and of the reasons for such action.

(2) The order required by paragraph (1)(A) or (B) of this subsection shall be issued within ninety days after the date of filing of the petition, except that the Secretary may (prior to such ninetieth day), by written notice to the petitioner, extend such ninety-day period to such time (not more than one hundred and eighty days after the date of filing of the petition) as the Secretary deems necessary to enable him to study and investigate the petition.

(3) No such regulation shall issue if a fair evaluation of the data before the Secretary–

(A) fails to establish that the proposed use of the food additive, under the conditions of use to be specified in the regulation, will be safe: *Provided,* That no additive shall be deemed to be safe if it is found to induce cancer when ingested by man or animal, or if it is found, after tests which are appropriate for the evaluation of the safety of food additives, to induce cancer in man or animal, except that this proviso shall not apply with respect to the use of a substance as an ingredient of feed for animals which are raised for food production, if the Secretary finds (i) that, under the conditions of use and feeding specified in proposed labeling and reasonably certain to be followed in practice, such additive will not adversely affect the animals for which such feed is intended, and (ii) that no residue of the additive will be found (by methods of examination prescribed or approved by the Secretary by regulations, which regulations shall not be subject to subsections (f) and (g)) in any edible portion of such animal after slaughter or in any food yielded by or derived from the living animal; or

(B) shows that the proposed use of the additive would promote deception of the consumer in violation of this Act or would otherwise result in adulteration or in misbranding of food within the meaning of this Act.

(4) If, in the judgement of the Secretary, based upon a fair evaluation of the data before him, a tolerance limitation is required in order to assure that the proposed use of an additive will be safe, the Secretary–

(A) shall not fix such tolerance limitation at a level higher than he finds to be reasonably required to accomplish the physical or other technical effect for which such additive is intended; and

(B) shall not establish a regulation for such proposed use if he finds upon a fair evaluation of the data before him that such data do not establish that such use would accomplish the intended physical or other technical effect.

(5) In determining, for the purpose of this section, whether a proposed use of a food additive is safe, the Secretary shall consider among other relevant factors–

(A) the probable consumption of the additive and of any substance formed in or on food because of the use of the additive;

(B) the cumulative effect of such additive in the diet of man or animals, taking into account any chemically or pharmacologically related substance or substances in such diet; and

(C) safety factors which in the opinion of experts qualified by scientific training and experience to evaluate the safety of food additives are generally recognized as appropriate for the use of animal experimentation data.

Regulation Issued on Secretary's Initiative

(d) The Secretary may at any time, upon his own initiative, propose the issuance of a regulation prescribing, with respect to any particular use of a food additive, the conditions under which such additive may be safely used, and the reasons therefor. After the thirtieth day following publication of such a proposal, the Secretary may by order establish a regulation based upon the proposal.

Publication and Effective Date of Orders

(e) Any order, including any regulation established by such order, issued under subsection (c) or (d) of this section, shall be published and shall be effective upon publication, but the Secretary may stay such effectiveness if, after issuance of such order, a hearing is sought with respect to such order pursuant to subsection (f).

(f)–

(1) Within thirty days after publication of an order made pursuant to subsection (c) or (d) of this section, any person adversely affected by such an order may file objections thereto with the Secretary, specifying with particularity the provisions of the order deemed objectionable, stating reasonable grounds therefor, and requesting a public hearing upon such objections. The Secretary shall, after due notice, as promptly as possible hold such public hearing for the purpose of receiving evidence relevant and material to the issues raised by such objections. As soon as practicable after completion of the hearing, the Secretary shall by order act upon such objections and make such order public.

(2) Such order shall be based upon a fair evaluation of the entire record at such hearing, and shall include a statement setting forth in detail the findings and conclusions upon which the order is based.

(3) The Secretary shall specify in the order the date on which it shall take effect, except that it shall not be made to take effect prior to the ninetieth day after its

publication, unless the Secretary finds that emergency conditions exist necessitating an earlier effective date, in which event the Secretary shall specify in the order his findings as to such conditions.

Judicial Review

(g)-

(1) In a case of actual controversy as to the validity of any order issued under subsection (f), including any order thereunder with respect to amendment or repeal of a regulation issued under this section, any person who will be adversely affected by such order may obtain judicial review by filing in the United States Court of Appeals for the circuit wherein such person resides or has his principal place of business, or in the United States Court of Appeals for the District of Columbia Circuit, within sixty days after the entry of such order, a petition praying that the order be set aside in whole or in part.

(2) A copy of such petition shall be forthwith transmitted by the clerk of the court to the Secretary, or any officer designated by him for that purpose, and thereupon the Secretary shall file in the court the record of the proceedings on which he based his order, as provided in section 2112 of title 28, United States Code. Upon the filing of such petition the court shall have jurisdiction, which upon the filing of the record with it shall be exclusive, to affirm or set aside the order complained of in whole or in part. Until the filing of the record the Secretary may modify or set aside his order. The findings of the Secretary with request to questions of fact shall be sustained if based upon a fair evaluation of the entire record at such hearing. The court shall advance on the docket and expedite the disposition of all causes filed therein pursuant to this section.

(3) The court, on such judicial review, shall not sustain the order of the Secretary if he failed to comply with any requirement imposed on him by subsection (f)(2) of this section.

(4) If application is made to the court for leave to adduce additional evidence, the court may order such additional evidence to be taken before the Secretary and to be adduced upon the hearing in such manner and upon such terms and conditions as to the court may seem proper, if such evidence is material and there were reasonable grounds for failure to adduce such evidence in the proceedings below. The Secretary may modify his findings as to the facts and order by reason of the additional evidence so taken, and shall file with the court such modified findings and order.

(5) The judgement of the court affirming or setting aside, in whole or in part, any order under this section shall be final, subject to review by the Supreme Court of the United States upon certiorari or certification as provided in section 1254 of title 28 of the United States Code. The commencement of proceedings under this section shall not, unless specifically ordered by the court to the contrary, operate as a stay of an order.

Amendment or Repeal of Regulations

(h) The Secretary shall by regulation prescribe the procedure by which regulations under the foregoing provisions of this section may be amended or repealed, and such procedure shall conform to the procedure provided in this section for the promulgation of such regulations.

Exemptions for Investigational Use

(i) Without regard to subsections (b) to (h), inclusive, of this section, the Secretary shall by regulation provide for exempting from the requirements of this section any food additive, and any food bearing or containing such additive, intended solely for investigational use by qualified experts when in his opinion such exemption is consistent with the public health.

BOTTLED DRINKING WATER

SEC. 410 [349][28] Whenever the Administrator of the Environmental Protection Agency prescribes interim or revised national primary drinking water regulations under section 1412 of the Public Health Service Act, the Secretary shall consult with the Administrator and within 180 days after the promulgation of such drinking water regulations either promulgate amendments to regulations under this chapter applicable to bottled drinking water or publish in the Federal Register his reasons for not making such amendments.

VITAMINS AND MINERALS

SEC. 411. [350][29]

(a)(1) Except as provided in paragraph (2)

(A) the Secretary may not establish, under section 201 (n), 401, or 403, maximum limits on the potency of any synthetic or natural vitamin or mineral within a food to which this section applies;

(B) the Secretary may not classify any natural or synthetic vitamin or mineral (or combination thereof) as a drug solely because it exceeds the level of potency which the Secretary determines is nutritionally rational or useful;

(C) the Secretary may not limit, under section 201(n), 401, or 403, the combination or number of any synthetic or natural–

(i) vitamin

(ii) mineral, or

(iii) other ingredient of food,

within a food to which this section applies.

(2) Paragraph (1) shall not apply in the case of a vitamin, mineral, other ingredient of food, or food, which is represented for use by individuals in the treatment or management of specific diseases or disorders, by children,

---

[28] Sec. 410 added by P.L. 93-523.
[29] Sec. 411 added by sec. 501(a) of P.L. 94-278.

or by pregnant or lactating women. For purposes of this subparagraph, the term "children" means individuals who are under the age of twelve years.

(b)(1) A food to which this section applies shall not be deemed under section 403 to be misbranded solely because its label bears, in accordance with section 403(i)(2), all the ingredients in the food or its advertising contains references to ingredients in the food which are not vitamins or minerals.

(2)(A) The labeling for any food to which this section applies may not list its ingredients which are not vitamins or minerals (i) except as a part of a list of all the ingredients of such food, and (ii) unless such ingredients are listed in accordance with applicable regulations under section 403. To the extent that compliance with clause (i) of this subparagraph is impracticable or results in deception or unfair competition, exemptions shall be established by regulations promulgated by the Secretary.

(B) Notwithstanding the provisions of subparagraph (A), the labeling and advertising for any food to which this section applies may not give prominence to or emphasize ingredients which are not:

(i) vitamins,

(ii) minerals, or

(iii) represented as a source of vitamins or minerals.

(c)(1) For purposes of this section, the term "food to which this section applies" means a food for humans which is a food for special dietary use-

(A) which is or contains any natural or synthetic vitamin or mineral, and

(B) which-

(i) is intended for ingestion in tablet, capsule, or liquid form, or

(ii) if not intended for ingestion in such a form, does not simulate and is not represented as conventional food and is not represented for use as a sole item of a meal or of the diet.

(2) For purposes of paragraph (l)(B)(i), a food shall be considered as intended for ingestion in liquid form only if it is formulated in a fluid carrier and it is intended for

ingestion in daily quantities measured in drops or similar small units of measure.

(3) For purposes of paragraph (1) and of section 403(j) insofar as that section is applicable to food to which this section applies, the term "special dietary use" as applied to food used by man means a particular use for which a food purports or is represented to be used, including but not limited to the following:

(A) Supplying a special dietary need that exists by reason of a physical, physiological, pathological, or other condition, including but not limited to the condition of disease, convalescence, pregnancy, lactation, infancy, allergic hypersensitivity to food, underweight, overweight, or the need to control the intake of sodium.

(B) Supplying a vitamin, mineral, or other ingredient for use by man to supplement his diet by increasing the total dietary intake.

(C) Supplying a special dietary need by reason of being a food for use as the sole item of the diet.

REQUIREMENTS FOR INFANT FORMULAS

SEC. 412.[350a][30]

(a)(1) An infant formula shall be deemed to be adulterated if–

(A) such infant formula does not provide nutrients as required by subsection (g);

(B) such infant formula does not meet the quality factor requirements prescribed by the Secretary under this section; or

(C) the processing of such infant formula is not in compliance with the quality control requirements prescribed by the Secretary under this section.

(2) The Secretary may by regulation–

(A) revise the list of nutrients in the table in subsection (g);

[30] Sec. 412 added by P.L. 96-359, September 26, 1980.

(B) revise the required level for any nutrient required by subsection (g);

(C) establish requirements for quality factors for such nutrients; and

(D) establish such quality control procedures as the Secretary determines necessary to assure that an infant formula provides nutrients in accordance with this section and establish requirements respecting the retention of records of procedures required under this clause (including maintaining necessary nutrient test records). Quality control procedures prescribed by the Secretary shall include the periodic testing of infant formulas to determine whether they are in compliance with this section.

(b)(1) On the ninetieth day after the date of the enactment of this section, and on each ninetieth day thereafter, a manufacturer of infant formula shall notify the Secretary that each infant formula manufactured by such manufacturer provides the nutrients required under subsection (g). Such notification requirement shall expire upon the effective date of regulations relating to quality control procedures prescribed by the Secretary under subsection (a)(2)(D).

(2) Not later than the ninetieth day before the first processing of any infant formula for commercial or charitable distribution for human consumption, the manufacturer shall notify the Secretary whether–

(A) such infant formula provides nutrients in accordance with subsection (g) and meets the quality factor requirements prescribed by the Secretary; and

(B) the processing of such infant formula meets the quality control procedure requirements prescribed by the Secretary.

(3) Before the first processing of any infant formula for commercial or charitable distribution for human consumption

(A) after a change in its formulation, or

(B) after a change in its processing,

which the manufacturer reasonably determines may affect whether the formula is adulterated as determined under subsection (a)(1), the manufacturer shall notify the Secretary of such changes and that the formula provides nutrients in accordance with subsection (g) and meets the quality factor requirements prescribed by the Secretary and that the processing of such infant formula is in compliance with the quality control procedures prescribed by the Secretary.

(c)(1) If the manufacturer of an infant formula has knowledge which reasonably supports the conclusion that an infant formula which has been processed by the manufacturer and which has left an establishment subject to the control of the manufacturer–

(A) may not provide the nutrients required by subsection (g); or

(B)(i) may be otherwise adulterated or misbranded, and (ii) if so adulterated or misbranded presents a risk to human health, the manufacturer shall promptly notify the Secretary of such non-compliance or risk to health.

(2) For purposes of paragraph (1), the term "knowledge" as applied to a manufacturer means (A) the actual knowledge that the manufacturer had, or (B) the knowledge which a reasonable person would have had under like circumstances or which would have been obtained upon the exercise of due care.

(d)(1) If a recall of an infant formula is begun by a manufacturer, the recall shall be carried out in accordance with such requirements as the Secretary may prescribe under paragraph (2), and–

(A) the Secretary shall, not later than the 15th day after the beginning of such recall and at least once every 15 days thereafter until the recall is terminated, review the actions taken under the recall to determine whether the recall meets the requirements prescribed under paragraph (2); and–

(B) the manufacturer shall, not later than the 14th day after the beginning of such recall and at least once every 14 days thereafter until the recall is terminated, report to the Secretary the actions taken to implement the recall.

(2) The Secretary shall by regulation prescribe the scope and extent of recalls of infant formulas necessary and appropriate for the degree of risk to human health presented by the formula subject to the recall.

(e)(1) Each manufacturer of an infant formula shall make and retain such records respecting the distribution of the infant formula through any establishment owned or operated by such manufacturer as may be necessary to effect and monitor recalls of the formula. No manufacturer shall be required under this subsection to retain any record respecting the distribution of an infant formula for a period of longer than 2 years from the date the record was made.

(2) To the extent that the Secretary determines that records are not being made or maintained in accordance with paragraph (1), the Secretary may by regulation prescribe the records required to be made under paragraph (1) and requirements respecting the retention of such records under such paragraph. Such regulations shall take effect on such date as the Secretary prescribes but not sooner than the 180th day after the date such regulations are promulgated. Such regulations shall apply only with respect to distributions of infant formulas made after such effective date.

(f)(1) Any infant formula which is represented and labeled for use by an infant–

(A) who has an inborn error of metabolism or a low birth weight, or

(B) who otherwise has an unusual medical or dietary problem, is exempt from the requirements of subsections (a) and (b). The manufacturer of an infant formula exempt under this paragraph shall, in the case of the exempt formula, be required to provide the notice required by subsection (c)(1) only with respect to adulteration or misbranding described in subsection (c)(1)(B), and to comply with the regulations prescribed by the Secretary under paragraph (2).

(2) The Secretary may by regulation establish terms and conditions for the exemption of an infant formula from

the requirements of subsections (a) and (b). An exemption of an infant formula under paragraph (1) may be withdrawn by the Secretary if such formula is not in compliance with applicable terms and conditions prescribed under this paragraph.

(g) An infant formula shall contain nutrients in accordance with the table set out in this subsection, or, if revised by the Secretary under subsection (a)(2), and so revised: (see table next page)

NUTRIENTS

Nutrient	Minimum[a]	Maximum[a]
Protein (gm)	1.8[b]	4.5
Fat:		
gm	3.3	6.0
percent cal	30.0	54.0
Essential fatty acids (linoleate):		
percent cal	2.7	
mg	300.0	
Vitamins:		
(A)(IU)	250.0 (75μg)[c]	750.0 (225 μg).[c]
D(IU)	40.0	100.0
K(μg)	4.0	
E(IU)	.7 (with 0.7 IU/gm linoleic acid)	
C (ascorbic acid)(mg)	8.0	
B_1(thiamine)(μg)	40.0	
B_2(riboflavin)(μg)	60.0	
B_6(pyridoxine)(μg)	35.0 (with 0.7 μg/gm of protein on formula).	
B_{12}(μg)	0.15	
Niacin(μg)	250.0	
Folic acid(μg)	4.0	
Pantothenic acid (μg)	300.0	
Biotin (μg)	1.5[d]	
Choline (mg)	7.0[d]	
Inositol (mg)	4.0[d]	
Minerals:		
Calcium (mg)	50.0[e]	
Phosphorus (mg)	25.0[e]	
Magnesium (mg)	6.0	
Iron (mg)	0.15	
Iodine (μg)	5.0	
Zinc (mg)	0.5	
Copper (μg)	60.0	
Manganese (μg)	5.0	
Sodium (mg)	20.0	60.0
Potassium (mg)	80.0	200.0
Chloride (mg)	55.0	150.0

[a] Stated per 100 kilocalories.
[b] The source of protein shall be at least nutritionally equivalent to casein.
[c] Retinol equivalents.
[d] Required to be included in this amount only in formulas which are not milk-based.
[e] Calcium to phosphorus ratio must be no less than 1.1 nor more than 2.0.

SECTION VII - GENERAL ADMINISTRATIVE PROVISIONS

Regulations And Hearings

SEC 701.[371]

(a) The authority to promulgate regulations for the efficient enforcement of this Act, except as otherwise provided in this section, is hereby vested in the Secretary.

(b) The Secretary of the Treasury and the Secretary of Health and Human Services shall jointly prescribe regulations for the efficient enforcement of the provisions of section 801, except as otherwise provided therein. Such regulations shall be promulgated in such manner and take effect at such time, after due notice, as the Secretary of Health and Human Services shall determine.

(c) Hearings authorized or required by this Act shall be conducted by the Secretary or such officer or employee as he may designate for the purpose.

(d) The definitions and standards of identity promulgated in accordance with the provisions of this Act shall be effective for the purposes of the enforcement of this Act, notwithstanding such definitions and standards as may be contained in other laws of the United States and regulations promulgated thereunder.

(e)(1) Any action for the issuance, amendment, or repeal of any regulation under section 401, 403(j), 404(a), 406, 501(b), or 502(d) or (h) of this Act shall be begun by a proposal made (A) by the Secretary on his own initiative, or (B) by petition of any interested persons, showing reasonable grounds therefor, filed with the Secretary. The Secretary shall publish such proposal and shall afford all interested persons an opportunity to present their views thereon, orally or in writing. As soon as practicable thereafter, the Secretary shall by order act upon such proposal and shall make such order public. Except as provided in paragraph (2), the order shall become effective at such time as may be specified therein, but not prior to the day following the last day on which objections may be field under such paragraph.

(2) On or before the thirtieth day after the date on which an order entered under paragraph (1) is made

public, any person who will be adversely affected by such order if placed in effect may file objections thereto with the Secretary, specifying with particularity the provisions of the order deemed objectionable, stating the grounds therefor, and requesting a public hearing upon such objections. Until final action upon such objections is taken by the Secretary under paragraph (3), the filing of such objections shall operate to stay the effectiveness of those provisions of the order to which the objections are made. As soon as practicable after the time for filing objections has expired the Secretary shall publish a notice in the Federal Register specifying those parts of the order which have been stayed by the filing of objections and, if no objections have been filed, stating that fact.

(3) As soon as practicable after such request for a public hearing, the Secretary, after due notice, shall hold such a public hearing for the purpose of receiving evidence relevant and material to the issues raised by such objections. At the hearing, any interested person may be heard in person or by representative. As soon as practicable after completion of the hearing, the Secretary shall by order act upon such objections and make such order public. Such order shall be based only on substantial evidence of record at such hearing and shall set forth, as part of the order, detailed findings of fact on which the order is based. The Secretary shall specify in the order the date on which it shall take effect, except that it shall not be made to take effect prior to the ninetieth day after its publication unless the Secretary finds that emergency conditions exist necessitating an earlier effective date, in which event the Secretary shall specify in the order his findings as to such conditions.

(f)(1) In a case of actual controversy as to the validity of any order under subsection (3), any person who will be adversely affected by such order if placed in effect may at any time prior to the ninetieth day after such order is issued file a petition with the Circuit Court of Appeals of the United States for the circuit wherein such person resides or has his principal place of business, for a judicial review of such order. A copy of the petition shall be forthwith transmitted by the clerk of the court to the Secretary or other officer designated by him for that purpose. The Secretary thereupon shall file in the court, the record of the proceedings on which the Secretary based his order, as provided in section 2112 of title 28, United States Code.

(2) If the petitioner applies to the court for leave to adduce additional evidence, and shows to this satisfaction of the court that such additional evidence is material and that there were reasonable grounds for the failure to adduce such evidence in the proceedings before the Secretary the court may order such additional evidence (and evidence in rebuttal thereof) to be taken before the Secretary, and to be adduced upon the hearing, in such manner and upon such terms and conditions as to the court may seem proper. The Secretary may modify his findings as to the facts, or make new findings, by reason of the additional evidence, so taken, and he shall file such modified or new findings, and his recommendation, if any, for the modification or setting aside of his original order, with the return of such additional evidence.

(3) Upon the filing of the petition referred to in paragraph (1) of this subsection, the court shall have jurisdiction to affirm the order, or to set it aside in whole or in part, temporarily or permanently. If the order of the Secretary refuses to issue, amend, or repeal a regulation and such order is not in accordance with law the court shall by its judgement order the Secretary to take action with respect to such regulations, in accordance with law. The findings of the Secretary as to the facts, if supported by substantial evidence, shall be conclusive (now covered by U.S.C. title 28, sec. 1254).

(4) The judgement of the court affirming or setting aside, in whole or in part, any such order of the Secretary shall be final, subject to review by the Supreme Court of the United States upon certiorari or certification as provided in sections 239 and 240 of the Judicial Code, as amended.

(5) Any action instituted under this subsection shall survive notwithstanding any change in the person occupying the office of Secretary or any vacancy in such office.

(6) The remedies provided for in this subsection shall be in addition to and not in substitution for any other remedies provided by law.

(g) A certified copy of the transcript of the record and proceedings under subsection (e) shall be furnished by the Secretary to any interested party at his request, and payment of the costs thereof, and shall be admissible in any criminal, libel for condemnation, exclusion of imports, or other proceedings arising under or in respect of this Act, irrespective of whether proceedings with respect to the order have previously been instituted or become final under subsection (f).

EXAMINATIONS AND INVESTIGATIONS

SEC. 702. [372]

(a) The Secretary is authorized to conduct examinations and investigations for the purpose of this Act through officers and employees of the Department or through any health, food or drug officer or employee of any State, Territory, or political subdivision thereof, duly commissioned by the Secretary as an officer of the Department. In the case of food packed in the Commonwealth of Puerto Rico or a Territory the Secretary shall attempt to make inspection of such food at the first point of entry within the United States, when in his opinion and with due regard to the enforcement of all the provisions of this Act, the facilities at his disposal will permit of such inspection. For the purposes of this subsection the term "United States" means the States and the District of Columbia.

(b) Where a sample of a food, drug, or cosmetic is collected for analysis under this Act the Secretary shall, upon request, provide a part of such official sample for examination or analysis by any person named on the label of the article, or the owner thereof, or his attorney or agent; except that the Secretary is authorized, by regulations, to make such reasonable exceptions from, and impose such reasonable terms and conditions relating to, the operation of this subsection as he finds necessary for the proper administration of the provisions of this Act.

(c) For purposes of enforcement of this Act, records of any department or independent establishment in the executive branch of the Government shall be open to inspection by any official of the Department of Health and Human Services duly authorized by the Secretary to make such inspection.

(d) The Secretary is authorized and directed, upon request from the Commissioner of Patents, to furnish full and complete information with respect to such questions relating to drugs as the Commissioner may submit concerning any patent application. The Secretary is further authorized, upon receipt of any such request, to conduct or cause to be conducted, such research as may be required.

(e)[57] [58] Any officer or employee of the Department designated by the Secretary to conduct examinations, investigations, or inspections under this Act relating to counterfeit drugs may, when so authorized by the Secretary–

(1) carry firearms;

(2) execute and serve warrants and arrest warrants;

(3) execute seizure by process issued pursuant to libel under section 304;

(4) make arrests without warrant for offenses under this Act with respect to such drugs if the offense is committed in his presence or, in the case of a felony, if he has probable cause to believe that the person so arrested has committed, or is committing, such offense; and

(5) make, prior to the institution of libel proceedings under section 304(a)(2), seizures of drugs or containers or of equipment, punches, dies, plates, stones, labeling, or other things, if they are, or he has reasonable grounds to believe that they are, subject to seizure and condemnation under such section 304(a)(2). In the event of seizure

---

[57] Section 1114 of title 18 of the United States Code is amended by sec. 17(h)(1) of P.L. 91-596. Sec. 1114 of title 18 will read in part "or any officer or employee of the Department of Health and Human Services or of the Department of Labor assigned to perform investigative, inspection or law enforcement functions."

[58] Subsec. (e) amended by sec. 701 of P.L. 91-513.

pursuant to this paragraph (5), libel proceedings under section 304(a)(2) shall be instituted promptly and the property seized be placed under the jurisdiction of the court.

SEAFOOD INSPECTION[59]

SEC. 702a.[372a]

The Secretary of Health and Human Services, upon application of any packer of any seafood for shipment or sale within the jurisdiction of this Act, may, at his discretion, designate inspectors to examine and inspect such food and the production, packing, and labeling thereof. If on such examination and inspection compliance is found with the provisions of this Act and regulations promulgated thereunder, the applicant shall be authorized or required to mark the food as provided by regulation to show such compliance. Services under this section shall be rendered only upon payment by the applicant of fees fixed by regulation in such amounts as may be necessary to provide, equip, and maintain an adequate and efficient inspection service. Receipts from such fees shall be covered into the Treasury and shall be available to the Secretary of Health and Human Services for expenditures incurred in carrying out the purposes of this section, including expenditures for salaries of additional inspectors when necessary to supplement the number of inspectors for whose salaries Congress has appropriated. The Secretary is hereby authorized to promulgate regulations governing the sanitary and other conditions which the services herein provided shall be granted and maintained and for otherwise carrying out the purposes of this section. Any person who forges, counterfeits, simulates, or falsely represents, or without proper authority uses any mark, stamp, tag, label, or other identification devices authorized or required by the provisions of this section or regulations thereunder, shall be guilty of a misdemeanor, and shall on conviction thereof be

[59] Sec. 902(a) provides that the amendment to the Food and Drug Act, section 10A, shall remain in force and effect and be applicable to the provisions of this Act. The Labor-Federal Security Appropriation Act of July 12, 1943 (ch. 221, title 11, §1,57 State. 500), renumbered this section as 702A of the Federal Food, Drug, and Cosmetic Act. Title 21 U.S.C., 1946 ed., codified this section as 372a.

subject to imprisonment for not more than one year or a fine of not less than $1,000 nor more than $5,000 or both such imprisonment and fine.

RECORDS OF INTERSTATE SHIPMENT

SEC. 703.[373][60]

For the purpose of enforcing the provisions of this Act, carriers engaged in interstate commerce, and persons receiving foods, drugs, devices, or cosmetics in interstate commerce or holding such articles so received, shall, upon the request of an officer or employee duly designated by the Secretary, permit such officer or employee, at reasonable times, to have access to and to copy all records showing the movement in interstate commerce of any food, drug, device, or cosmetic, or the holding thereof during or after such movement, and the quantity, shipper, and consignee thereof; and it shall be unlawful for any such carrier or person to fail to permit such access to and copying of any such record so requested when such request is accompanied by a statement in writing specifying the nature or kind of food, drug, device, or cosmetic to which such request relates: *Provided,* That evidence obtained under this section, or any evidence which is directly or indirectly derived from such evidence, shall not be used in a criminal prosecution of the person from whom obtained: *Provided further,* That carriers shall not be subject to the other provisions of this Act by reason of their receipt, carriage, holding, or delivery of food, drugs, devices, or cosmetics in the usual course of business as carriers.

FACTORY INSPECTION

SEC. 704.[374][61]

(a)(1) For purposes of enforcement of this Chapter, officers or employees duly designated by the Secretary, upon presenting appropriate credentials and a written notice to the owner, oper-

[60] Sec. 703 amended by sec. 230 of P.L. 91-452.
[61] Sec. 704 amended by sec. 6 of P.L. 94-295.

ator, or agent in charge, are authorized (A) to enter, at reasonable times, any factory, warehouse, or establishment in which food, drugs, devices, or cosmetics are manufactured, processed, packed, or held, for introduction into interstate commerce or after such introduction, or to enter any vehicle being used to transport or hold such food, drugs, devices, or cosmetics in interstate commerce; and (B) to inspect, at reasonable times and within reasonable limits and in a reasonable manner, such factory, warehouse, establishment, or vehicle and all pertinent equipment, finished and unfinished materials, containers, and labeling therein. In the case of any factory, warehouse, establishment, or consulting laboratory in which prescription drugs or restricted devices are manufactured, processed, packed, or held, the inspection shall extend to all things therein (including records, files, papers, processes, controls, and facilities) bearing on whether prescription drugs or restricted devices which are adulterated or misbranded within the meaning of this Chapter, or which may not be manufactured, introduced into interstate commerce, or sold, or offered for sale by reason of any provision of this Chapter, have been or are being manufactured, processed, packed, transported, or held in any such place, or otherwise bearing on violation of this Chapter. No inspection authorized by the proceeding sentence or by paragraph (3) shall extend to financial data, sales data other than shipment data, pricing data, personnel data (other than data as to qualifications of technical and professional personnel performing functions subject to this Chapter), and research data (other than data relating to new drugs, antibiotic drugs, and devices and subject to reporting and inspection under regulations lawfully issues pursuant to section 505 (i) or (k), section 507 (d) or (g), section 519, or 520 (g), and data relating to other drugs or devices which in the case of a new drug would be subject to reporting or inspection under lawful regulations issued pursuant to section 505 (k) of the title). A separate notice shall be given for each such inspection, but a notice shall not be required to each entry made during the period covered by the inspection. Each such inspection shall be commenced and completed with reasonable promptness.

(2) The provisions of the second sentence of this subsection shall not apply to–

(A) pharmacies which maintain establishments in conformance with any applicable local laws regulating the practice of pharmacy and medicine and which are regularly engaged in dispensing prescription drugs, or devices upon prescriptions of practitioners licensed to administer such drugs or devices to patients under the care of such practitioners in the course of their professional practice, and which do not, either through a subsidiary or otherwise, manufacture, prepare, propagate, compound, or process drugs or devices for sale other than in the regular course of their business of dispensing or selling drugs or devices at retail;

(B) practitioners licensed by law to prescribe or administer drugs or prescribe or use devices, as the case may be, and who manufacture, prepare, propagate, compound, or process drugs or manufacture or process devices solely for use in the course of their professional practice;

(C) persons who manufacture, prepare, propagate, compound, or process drugs or manufacture or process devices solely for use in research, teaching, or chemical analysis and not for sale;

(D) such other classes of persons as the Secretary may by regulation exempt from the application of this section upon a finding that inspection as applied to such classes of persons in accordance with this section is not necessary for the protection of the public health.

An officer or employee making an inspection under paragraph (1) for purposes of enforcing the requirements of section 412 of this title applicable to infant formulas shall be permitted, at all reasonable times, to have access to and to copy and verify any records–

(A) bearing on whether the infant formula manufactured or held in the facility inspected meets the requirements of section 412 of this title, or

(B) required to be maintained under section 412 of this title.

(b) Upon completion of any such inspection of a factory, warehouse, consulting laboratory, or other establishment, and prior to leaving the premises, the officer or employee making the inspection shall give to the owner, operator, or agent in charge a report in writing setting forth any conditions or practices observed by him which, in his judgement, indicate that any food, drug, device, or cosmetic in such establishment

(1) consists in whole or in part of any filthy, putrid, or decomposed substance, or

(2) has been prepared, packed, or held under insanitary conditions whereby it may have become contaminated with filth, or whereby it may have been rendered injurious to health. A copy of such report shall be sent promptly to the Secretary.

(c) If the officer or employee making any such inspection of a factory, warehouse, or other establishment has obtained any sample in the course of the inspection, upon completion of the inspection and prior to leaving the premises he shall give to the owner, operator, or agent in charge a receipt describing the samples obtained.

(d) Whenever in the course of any such inspection of a factory or other establishment where food is manufactured, processed, or packed, the officer or employee making the inspection obtains a sample for the purpose of ascertaining whether such food consists in whole or in part of any filthy, putrid, or decomposed substance, or is otherwise unfit for food, a copy of the results of such analysis shall be furnished promptly to the owner, operator, or agent in charge.

(e)[62] Every person required under section 519 or 520(g) to maintain records and every person who is in charge or custody of such records shall, upon request of an officer or employee designated by the Secretary, permit such officer or employee at all reasonable times to have access to and to copy and verify such records.

[62] Subsec. 704(e) added by sec. 6(d) of P.L. 94-295.

PUBLICITY

SEC. 705.[375]

(a) The Secretary shall cause to be published from time to time reports summarizing all judgements, decrees, and court orders which have been rendered under this Act, including the nature of the charge and the disposition thereof.

(b) The Secretary may also cause to be disseminated information regarding food, drugs, devices, or cosmetics in situations involving, in the opinion of the Secretary, imminent danger to health, or gross deception of the consumer. Nothing in this section shall be construed to prohibit the Secretary from collecting, reporting, and illustrating the results of the investigations of the Department.

LISTING AND CERTIFICATION OF COLOR ADDITIVES FOR FOODS, DRUGS, AND COSMETICS

When Color Additives Deemed Unsafe

SEC. 706.[376][63]

(a) A color additive (for which it is being used or intended to be used or is represented as suitable) in or on food or drugs or devices or cosmetics be deemed unsafe for the purpose of the application of section 402(c), section 501(a)(4), or section 601(e), as the case may be unless–

(1)(A) there is in effect, and such additive and such use are in conformity with, a regulation issued under subsection (b) of this section listing such additive for such use, including any provision of such regulation prescribing the conditions under which such additive may be safely used, and (b) such additive either (i) is from a batch certified, in accordance with regulations issued pursuant to subsection (c), for such use, or (ii) has, with respect to such use, been exempted by the Secretary from the requirement of certification; or

---

[63] Sec. 706 amended by sec. 9(a) of P.L. 94-295.

(2) such additive and such use thereof conform to the terms of an exemption which is in effect pursuant to subsection (f) of this section.

While there are in effect regulations under subsections (b) and (c) of this section relating to a color additive or an exemption pursuant to subsection (f) with respect to such additive, an article shall not, by reason of bearing or containing such additive in all respects in accordance with such regulations or such exemption, be considered adulterated within the meaning of clause (1) of section 402(a) if such article is a food, or within the meaning of section 601(a) if such article is a cosmetic other than a hair dye (as defined in the last sentence of section 601(a)). A color additive for use in or on a device shall be subject to this section only if the color additive comes in direct contact with the body of man or other animals for a significant period of time. The Secretary may by regulation designate the use of color additives in or on devices which are subject to this section.

Listing of Colors

(b)(1) The Secretary shall, by regulation, provide for separately listing color additives for use in or on food, color additives for use in or on drugs or devices, and color additives for use in or on cosmetics. If and to the extent that such additives are suitable and safe for any such use when employed in accordance with such regulations.

(2)(A) Such regulations may list any color additive for use generally in or on food, or in or on drugs or devices, or in or on cosmetics, if the Secretary finds that such additive is suitable and may safely by employed for such general use.

(B) If the data before the Secretary do not establish that the additive satisfies the requirement for listing such additive on the applicable list pursuant to subparagraph (A) of this paragraph, or if the proposal is for listing such additive for a more limited use or uses, such regulations may list such additive only for any more limited use or uses for which it is suitable and may safely be employed.

(3) Such regulations shall, to the extent deemed necessary by the Secretary to assure the safety of the use or uses for which a particular color additive is listed, prescribe the conditions under which such additive may be safely employed for such use or uses (including, but not limited to, specifications, hereafter in this section referred to as tolerance limitations, as to the maximum quantity or quantities which may be used or permitted to remain in or on the article or articles in or on which it is used; specifications as to the manner in which such additive may be added to or used in or on such article or articles; and directions or other labeling or packaging requirements for such additive).

(4) The Secretary shall not list a color additive under this section for a proposed use unless the data before him establish that such use, under the conditions of use specified in the regulations, will be safe: *Provided,* however, That a color additive shall be deemed to be suitable and safe for the purpose of listing under this subsection for use generally in or on food, while there is in effect a published finding of the Secretary declaring such substance exempt from the term "food additive" because of its being generally recognized by qualified experts as safe for its intended use, as provided in section 201(s).

(5)(A) In determining, for the purpose of this section, whether a proposed use of a color additive is safe, the Secretary shall consider, among other relevant factors

(i) the probable consumption of, or other relevant exposure from, the additive and of any substance formed in or on food, drugs, devices, or cosmetics because of the use of the additive;
(ii) the cumulative effect, if any, of such additive in the diet of man or animals, taking into account the same or any chemically or pharmacologically related substance or substances in such diet.
(iii) safety factors which, in the opinion of experts qualified by scientific training and

experience to evaluate the safety of color additives for the use or uses for which the additive is proposed to be listed, are generally recognized as appropriate for the use of animal experimentation data; and

(iv) the availability of any needed practicable methods of analysis for determining the identity and quantity of (I) the pure dye and all intermediates and other impurities contained in such color additive, (II) such additive in or on any article of food, drug, device, or cosmetic, and (III) any substance formed in or on such article because of the use of such additive.

(B) A color additive (i) shall be deemed unsafe, and shall not be listed, for any use which will or may result in ingestion of all or part of such additive, if the additive is found by the Secretary to induce cancer when ingested by man or animal, or if it is found by the Secretary, after tests which are appropriate for the evaluation of the safety of additives for use in food, to induce cancer in man or animal, and (ii) shall be deemed unsafe, and shall not be listed, for any use which will not result in ingestion of any part of such additive, if, after tests which are appropriate for the evaluation of the safety of additives for such use, or after other relevant exposure of man or animal to such additive, it is found by the Secretary to induce cancer in man or animal: *Provided,* That clause (i) of this sub-paragraph (B) shall not apply with respect to the use of a color additive as an ingredient of feed for animals which are raised for food production, if the Secretary finds that, under the conditions of use and feeding specified in proposed labeling and reasonably certain to be followed in practice, such additive will not adversely affect the animals for which such feed is intended, and that no residue of the additive will be found by methods of examination prescribed or approved by the Secretary by regulations, which regulations shall not be subject to subsection (d) in any edible portion of such animals after slaughter or in any food yielded by or derived from the living animal.

(c)(i) In any proceeding for the issuance, amendment, or repeal of a regulation listing a color additive, whether commenced by a proposal of the Secretary on his own initiative or by a proposal contained in a petition, the petitioner or any other person who will be adversely affected by such proposal or by the Secretary's order issued in accordance with paragraph (1) of section 701(e) if placed in effect, may request, within the time specified in this subparagraph, that the petition or order thereon, or the Secretary's proposal, be referred to an advisory committee for a report and recommendations with respect to any matter arising under subparagraph (B) of this paragraph, which is involved in such proposal or order and which requires the exercise of scientific judgement. Upon such request, or if the Secretary within such time deems such a referral necessary, the Secretary shall forthwith appoint an advisory committee under subparagraph (C) of this paragraph and shall refer to it, together with all the data before him, such matter arising under sub-paragraph (B) for study thereof and for a report and recommendations on such matter. A person who has filed a petition or who has requested the referral of a matter to an advisory committee pursuant to this sub-paragraph (C), as well as representatives of the Department of Health and Human Services, shall have the right to consult with such advisory committee in connection with the matter referred to it. The request for referral under this subparagraph, or the Secretary's referral on his own initiative, may be made at any time before, or within thirty days after, publication of an order of the Secretary acting upon the petition or proposal.

(ii) Within sixty days after the date of such referral, or within an additional thirty days if the committee deems such additional time necessary, the committee shall, after independent study of the data furnished to it by the Secretary and other data before it, certify to the Secretary a report and recommendations, together with all underlying data and a statement of the reasons or basis for the recommendations. A copy of the foregoing shall be promptly supplied by the Secretary to any person who has filed a petition, or who has requested such referral to the advisory committee. Within thirty days after such certification, and after giving due consideration, underlying data, and statement, and to any prior order issued by him in connection with such matter, the Secretary shall by order confirm or modify any order therefore issued, or, if no such prior order has been issued, shall by order act upon the petition or other proposal.

(iii) Where–

(I) by reason of subparagraph (B) of this paragraph, the Secretary has initiated a proposal to remove from listing a color additive previously listed pursuant to this section; and

(II) a request has been made for referral of such proposal to an advisory committee;

the Secretary may not act by order on such proposal until the advisory committee has made a report and recommendations to him under clause (ii) of this subparagraph and he has considered such recommendations, unless the Secretary finds that emergency conditions exist necessitating the issuance of an order notwithstanding this clause.

(C)[64] The advisory committee referred to in subparagraph C) of this paragraph shall be composed of experts selected by the National Academy of Sciences, qualified in the subject matter referred to the committee and of adequately diversified professional background, except that in the event of the inability or refusal of the National Academy of Sciences to act, the Secretary shall select the members of the committee. The size of the committee shall be determined by the Secretary. Members of any advisory committee established under this Act, while attending conferences or meetings of their committees or otherwise serving at the request of the Secretary, shall be entitled to receive compensation at rates to be fixed by the Secretary but at rates not exceeding the daily equivalent of the rate specified at the time of such service for grade GS-18 of the General Schedule, including traveltime; and while away from

[64] Sec. 706(b)(5)(D) amended by sec. 601(d)(2) of P.L. 91-515.

their homes or regular places of business they may be allowed travel expenses, including per diem in lieu of subsistence as authorized by section 5703(b) of title 5 of the United States Code for persons in the Government service employed intermittently. The members shall not be subject to any other provisions of law regarding the appointment and compensation of employees of the United States. The Secretary shall furnish the committee with adequate clerical and other assistance, and shall by rules and regulations prescribe the procedure to be followed by the committee.

(6) The Secretary shall not list a color additive under this subsection for a proposed use if the data before him show that such proposed use would promote deception of the consumer in violation of this Act or would otherwise result in misbranding or adulteration within the meaning of this Act.

(7) If, in the judgment of the Secretary, a tolerance limitation is required in order to assure that a proposed use of a color additive will be safe, the Secretary

(A) shall not list the additive for such use if he finds that the data before him do not establish that such additive, if used within a safe tolerance limitation, would achieve the intended physical or other technical effect; and

(B) shall not fix such tolerance limitation at a level higher than he finds to be reasonably required to accomplish the intended physical or other technical effect.

(8) If, having regard to the aggregate quantity of color additive likely to be consumed in the diet or to be applied to the human body, the Secretary finds that the data before him fail to show that it would be safe and otherwise permissible to list a color additive (or pharmacologically related color additives) of all uses proposed therefor and at the levels of concentration proposed, the Secretary shall, in determining for which use or uses such additive (or such related additives) shall be or remain listed, or how the aggregate allowable safe tolerance for such additive or additives shall be allowed by him

among the uses under consideration, take into account, among other relevant factors (and subject to the paramount criterion of safety), (A) the relative marketability of the articles involved as affected by the proposed uses of the color additive (or of such related additives) in or on such articles, and the relative dependence of the industries concerned on such uses; (B) the relative aggregate amounts of such color additive which he estimates would be consumed in the diet or applied to the human body by reason of the various uses and levels of concentration proposed; and (C) the availability, if any, of other color additives suitable and safe for one or more of the uses proposed.

Certification of Colors

(c) The Secretary shall further, by regulation, provide (1) for the certification, with safe diluents or without diluents, of batches of color additives listed pursuant to subsection (b) and conforming to the requirements for such additives established by regulations under such subsection and this subsection, and (2) for exemption from the requirement of certification in the case of any such additive, or any listing or use thereof, for which he finds such requirement not to be necessary in the interest of the protection of the public health: *Provided,* That, with respect to any use in or on food for which a listed color additive is deemed to be safe by reason of the provision to paragraph (4) of subsection (b), the requirement of certification shall be deemed not to be necessary in the interest of public health protection.

Procedure of Issuance, Amendment, or Repeal of Regulations

(d) The provisions of section 701(e), (f), and (g) of this Act shall, subject to the provisions of subparagraph (C) of subsection (B) (5) of this section, apply to and in all respects govern proceedings for the issuance, amendment, or repeal of regulations under subsection (b) or (c) of this section (including judicial review of the Secretary's action in such proceedings) and the admissibility of transcripts of the record of such proceedings in other proceedings, except that–

(1) if the preceding is commenced by the filing of a petition, notice of the proposal made by the petition shall be published in general terms by the Secretary within thirty days after such filing and the Secretary's order (required by paragraph (1) of section 701(e) acting upon such proposal shall, in the absence of prior referral (or request for referral) to an advisory committee, be issued within ninety days after the date of such filing, except that the Secretary may (prior to such ninetieth day) by written notice to the petitioner, extend such ninety-day period to such time (not more than one hundred and eighty days after the date of filing of the petition) as the Secretary deems necessary to enable him to study and investigate the petition;

(2) any report, recommendations, underlying data, and reasons certified to the Secretary by an advisory committee appointed pursuant to subparagraph (D) of subsection (b) (5) of this section, shall be made a part of the record of any hearing if relevant and material, subject to the provisions of section 7(c) of the Administrative Procedure Act (5 U.S.C., sec. 1006(c). The advisory committee shall designate a member to appear and testify at any such hearing with respect to the report and recommendations of such committee upon request of the Secretary, the petitioner, or the officer conducting the hearing, but this shall not preclude any other member of the advisory committee from appearing and testifying at such hearing;

(3) the Secretary's order after public hearing (acting upon objections filed to an order made prior to hearings) shall be subject to the requirements of section 409(f)(2); and

(4) the scope of judicial review of such order shall be in accordance with the fourth sentence of paragraph (2), and with the provisions of paragraph (3), of section 409(g).

Fees

(e) The admitting to listing and certification of color additives, in accordance with regulations prescribed under this Act, shall be performed only upon payment of such fees, which shall be specified in such regulations, as may be necessary to provide, maintain, and equip an adequate service for such purposes.

Exemptions

(f) The Secretary shall by regulations issued without regard to subsection (d) provide for exempting from the requirements of this section any color additive or any specific type of use thereof, and any article of food, drug, device, or cosmetic bearing or containing such additive, intended solely for investigational use by qualified experts when in his opinion such exemption is consistent with the public health.

[The Color Additive Amendments to the Federal Food, Drug, and Cosmetic Act took effect on the date of enactment, July 12, 1960, subject to the provisions of sec. 203, title II, of P.L. 86-618 which follows:]

Provisional Listings of Commercially Established Colors

(a)(1) The purpose of this section is to make possible, on an interim basis for a reasonable period, through provisional listings, the use of commercially established color additives to the extent consistent with the public health, pending the completion of the scientific investigations needed as a basis for making determinations as to listing of such additives under [the basic Act the Federal Food, Drug, and Cosmetic Act] as amended by this Act.[65] A Provisional listing (including a deemed provisional listing) of a color additive under this section for any use shall, unless sooner terminated or expiring under the provisions of this section, expire (A) on the closing date (as defined in paragraph (2) of this subsection), or (B) on the effective date of a listing of such additive for such use under section 706 of the basic Act, whichever date first occurs.

[65] Words "this Act" refer to P.L. 86-618.

(2) For the purposes of this section, the term "closing date" means (A) the last day of the two and one-half year period beginning on the enactment date, or (B), with respect to a particular provisional listing (or deemed provisional listing) of a color additive or use thereof, such later closing date as the Secretary may from time to time establish pursuant to the authority of this paragraph. The Secretary may by regulation, upon application of an interested person or on his own initiative, from time to time postpone the original closing date with respect to a provisional listing (or deemed provisional listing) under this section of a specified color additive, or of a specified use or uses of such additive, for such period or periods as he finds necessary to carry out the purpose of this section, if in the Secretary's judgment such action is consistent with the objective of carrying to completion in good faith, as soon as reasonably practicable, the scientific investigations necessary for making a determination as to listing such additive, or such specified use or uses thereof, under section 706 of the basic Act. The Secretary may terminate a postponement of the closing date at any time if he finds that such postponement should not have been granted, or that by reason of a change in circumstances the basis for such postponement no longer exists, or that there has been a failure to comply with a requirement for submission of progress reports or with other conditions attached to such postponement.

(b) Subject to the other provisions of this section–

(1) any color additive which on the day preceding the enactment date, was listed and certifiable for any use or uses under section 406(b), 504, 604, or under the third proviso of section 402(c), of the basic Act, and of which a batch or batches had been certified for such use or uses prior to the enactment date, and

(2) any color additive which was commercially used or sold prior to the enactment date for any use or uses in or any food, drug, device, or cosmetic, and which either (A), on the day preceding the enactment date, was not a material within the purview of any of the provisions of

the basic Act enumerated in paragraph (1) of this subsection, or (B) is the color additive known as synthetic betacarotene, shall, beginning on the enactment date, be deemed to be provisionally listed under this section as a color additive for such use or uses.

(c) Upon request of any person, the Secretary, by regulations issued under subsection (d), shall without delay, if on the basis of the data before him he deems such action consistent with the protection of the public health, provisionally list a material as a color additive for any use for which it was listed, and for which a batch or batches of such material had been certified, under section 406(b), 504, or 604 of the basic Act prior to the enactment date, although such color was no longer listed and certifiable for such use under such sections on the day preceding the enactment date. Such provisional listing shall take effect on the date of publication.

(d)(1) The Secretary shall, by regulations issued or amended from time to time under this section–

(A) insofar as practicable promulgate and keep current a list or lists of the color additives, and of the particular uses thereof, which he finds are deemed provisionally listed under subsection (b), and the presence of a color additive on such a list with respect to a particular use shall, in any proceeding under the basic Act, be conclusive evidence that such provisional listing is in effect;

(B) provide for the provisional listing of the color additives and particular uses thereof specified in subsection (c);

(C) provide with respect to particular uses for which color additives are or are deemed to be provisionally listed, such temporary tolerance limitations (including such limitations at zero level) and other conditions of use and labeling or packaging requirements, if any, as in his judgement are necessary to protect the public health pending listing under section 706 of the basic Act;

(D) provide for the certification of batches of such color additives (with or without diluents) for the uses for which they are so listed or deemed to be listed under this section, except that such an additive which is a color additive deemed provisionally listed under subsection (b)(2) of this section shall be deemed exempt from the requirement of such certification while not subject to a tolerance limitation; and

(E) provide for the termination of a provisional listing (or deemed provisional listing) of a color additive or particular use thereof forthwith whenever in his judgement such action is necessary to protect the public health.

(2)(A) Except as provided in subparagraph (C) of this paragraph, regulations under this section shall, from time to time, be issued, amended, or repealed by the Secretary without regard to the requirements of the basic Act, but for the purposes of application of section 706(e) of the basic Act (relating to fees) and of determining the availability of appropriations of fees (and of advance deposits to cover fees), proceedings, regulations, and certifications under this section shall be deemed to be proceedings, regulations, and certifications under such section 706. Regulations providing for fees (and advance deposits to cover fees), which on the day preceding the enactment date were in effect pursuant to section 706 of the basic Act, shall be deemed to be regulations under section 706 as in effect prior to the enactment date shall be available for the purposes specified in such section 706 as so amended.

(B) If the Secretary, by regulation–

(i) has terminated a provisional listing (or deemed provisional listing) of a color additive or particular use thereof pursuant to paragraph (1)(E) of this subsection; or

(ii) has, pursuant to paragraph (1)(C) or paragraph (3) of this subsection, initially established or rendered more restrictive a tolerance limitation or other restriction or requirement with respect to a provisional listing (or deemed provisional listing) which listing had become effective prior to such action, any person adversely affected by such action may, prior to the expiration of the period specified in clause (A) of subsection (a)(2) of this section, file with the Secretary a petition for amendment of such regulation so as to revoke or modify such action of the Secretary, but the filing of such petition shall not operate to stay or suspend the effectiveness of such action. Such petition shall, in accordance with regulations, set forth the proposed amendment and shall contain data (or refer to data which are before the Secretary or of which he will take official notice), which show that the revocation or modification proposed is consistent with the protection of the public health. The Secretary shall, after publishing such proposal and affording all interested persons an opportunity to present their views thereon orally or in writing, act upon such proposal by published order.

(C) Any person adversely affected by an order entered under subparagraph (B) of this paragraph may, within thirty days after its publication, file objections thereto with the Secretary, specifying with particularity the provisions of the order deemed objectionable, stating reasonable grounds for such objections, and requesting a public hearing upon such objections. The Secretary shall hold a public hearing on such objections and shall, on the basis of the evidence adduced at such hearing, act on such objections by published order. Such order may reinstate a terminated provisional listing, or increase or dispense with a previously established temporary tolerance limitation, or make less restrictive any other limitation established by him under paragraph (1) or (3) of this subsection, only if in his judgement the evidence so adduced shows that such actions will be consistent with the protection of the public health. An order entered under this subparagraph shall be subject to judicial review in accordance with section 701(f) of the basic Act except that the findings and order of the Secretary shall be sustained only if based upon a fair evaluation of the entire record at such hearing. No stay or suspension of such order shall be ordered by the court pending conclusion of such judicial review.

(D) On and after the enactment date, regulations, provisional listings, and certifications (or exemptions from certification) in effect under this section shall, for the purpose of determining whether an article is adulterated or misbranded within the meaning of the basic Act by reason of its being, bearing or containing a color additive, have the same effect as would regulations, listings, and certifications (or exemptions from certification) under section 706 of the basic Act. A regulation, provisional listing or termination thereof, tolerance, limitation, or certification or exemption therefrom, under this section shall not be the basis for any presumption or inference in any preceding under section 706(b) or (c) of the basic Act.

(3) For the purpose of enabling the Secretary to carry out his functions under paragraphs (1)(A) and (C) of this subsection with respect to color additives deemed provisionally listed, he shall, as soon as practicable after enactment of this Act, afford by public notice a reasonable opportunity to interested persons to submit data relevant thereto. If the data so submitted or otherwise before him do not, in his judgement, establish a reliable basis for including such a color additive or particular use or uses thereof in a list or lists promulgated under paragraph (1)(A), or for determining the prevailing level or levels of use thereof prior to the enactment date with a view to prescribing a temporary tolerance or tolerances for such use or uses under paragraph (1)(C), the Secretary shall establish a temporary tolerance limitation at zero level for such use or uses until such time as he finds that

it would not be inconsistent with the protection of the public health to increase or dispense with such temporary tolerance limitation.

REVISION OF UNITED STATES PHARMACOPEIA: DEVELOPMENT OF ANALYSIS AND MECHANICAL AND PHYSICAL TESTS

[377][66] The Secretary, in carrying into effect the provisions of this chapter, is authorized hereafter to cooperate with associations and scientific societies in the revision of the United State Pharmacopeia and in the development of methods of analysis and mechanical and physical tests necessary to carry out the work of the Food and Drug Administration. [From the Labor-Federal Security Appropriation Act, 1944.]

SEC. 707.[67] [378]

(a)(1) Except as provided in subsection (c), before the Secretary may initiate any action under Chapter III.

(A) with respect to any food which the Secretary determines is misbranded under section 403(a)(2) because of its advertising, or

(B) with respect to a food's advertising which the Secretary determines causes the food to be so misbranded,

the Secretary shall, in accordance with paragraph (2), notify in writing the Federal Trade Commission of the action the Secretary proposes to take respecting such food or advertising.

(2) The notice required by paragraph (1) shall–

(A) contain

(i) a description of the action the Secretary proposes to take and of the advertising which

---

[66] Section 377, 21 U.S. Code is from the Labor-Federal Security Appropriations Act, 1944 (57 Stat. 500), not from the Federal Food, Drug, and Cosmetic Act. All functions of the Federal Security Administrator were transferred to the Secretary of Health and Human Services.

[67] Sec. 707 added by P.L. 94-278.

the Secretary has determined causes a food to be misbranded,

(ii) a statement of the reasons for the Secretary's determination that such advertising has caused such food to be misbranded, and

(B) be accompanied by the records, documents, and other written materials which the Secretary determines supports his determination that such food is misbranded because of such advertising.

(b)(1) If the Secretary notifies the Federal Trade Commission under subsection (a) of action proposed to be taken under Chapter III with respect to a food or food advertising and the Commission notifies the Secretary in writing, within the 30-day period beginning on the date of the receipt of such notice, that–

(A) it has initiated under the Federal Trade Commission Act an investigation of such advertising to determine if it is prohibited by such Act or any order or rule under such Act,

(B) it has commenced (or intends to commence) a civil action under section 5, 13 or 19 with respect to such advertising or the Attorney General has commenced (or intends to commence) a civil action under section 5 with respect to such advertising,

(C) it has issued and served (or intends to issue and serve) a complaint under section 5(b) of such Act respecting such advertising, or

(D) pursuant to section 16(b) of such Act it has made a certification to the Attorney General respecting such advertising, the Secretary may not, except as provided by paragraph (2), initiate the action described in the Secretary's notice to the Federal Trade Commission.

(2) If, before the expiration of the 60-day period beginning on the date the Secretary receives a notice described in paragraph (1) from the Federal Trade Commission in response to a notice of the Secretary under subsection (a)–

(A) the Commission or the Attorney General does not commence a civil action described in subparagraph (B) of paragraph (1) of this subsection respecting the advertising described in the Secretary's notice,

(B) the Commission does not issue and serve a complaint described in subparagraph (C) of such paragraph respecting such advertising, or

(C) the Commission does not (as described in subparagraph (D) of such paragraph) make a certification to the Attorney General respecting such advertising, or, if the Commission does make such a certification to the Attorney General respecting such advertising, the Attorney General, before the expiration of such period, does not cause appropriate criminal proceedings to be brought against such advertising, the Secretary may, after the expiration of such period, initiate the action described in the notice to the Commission pursuant to subsection (a). The Commission shall promptly notify the Secretary of the commencement by the Commission of such a civil action, the issuance and service by it of such a complaint, or the causing by the Attorney General of criminal proceedings to be brought against such advertising.

(c) The requirements of subsections (a) and (b) do not apply with respect to action under Chapter III with respect to any food or food advertising if the Secretary determines that such action is required to eliminate an imminent hazard to health.

(d) For the purpose of avoiding unnecessary duplication, the Secretary shall coordinate any action taken under Chapter III because of advertising which the Secretary determines causes a food to be misbranded with any action of the Federal Trade Commission under the Federal Trade Commission Act with respect to such advertising.

CONFIDENTIAL INFORMATION

SEC. 708.[68] [379] The Secretary may provide any information which is exempt from disclosure pursuant to subsection (a) of section 552 of the title 5, United States Code, by reason of subsection (b) (4) of such section to a person other than an officer or employee of the Department if the Secretary determines such other person requires the information in connection with an activity which is undertaken under contract with the Secretary, which relates to the Administration of this Act, and with respect to which the Secretary (or an officer or employee of the Department) is not prohibited from using such information. The Secretary shall require as a condition to the provision of information under this section that the person receiving it take such security precautions respecting the information as the Secretary may by regulation prescribe.

PRESUMPTION

SEC. 709.[379A] In any action to enforce the requirements of this Act respecting a device the connection with interstate commerce required for jurisdiction in such action shall be presumed to exist.

SECTION VIII - IMPORTS AND EXPORTS

SEC. 801.[69] [381]

(a) The Secretary of the Treasury shall deliver to the Secretary of Health and Human Services, upon his request, samples of food, drugs, devices, and cosmetics which are being imported or offered for import into the United States, giving notice thereof to the owner or consignee, who may appear before the Secretary of Health and Human Services and have the right to introduce testimony. The Secretary of Health and Human Services shall furnish to the Secretary of the Treasury a list of establishments registered pursuant to subsection (i) of section 510 and shall

---

[68] Secs. 708 and 709 added by sec. 8 of P.L. 94-295.

[69] Sec. 801 amended by sec. 701 of P.L. 91-513, and by secs. 3(f) and 4(b) (3) of P.L. 94-295.

request that if any drugs or devices manufactured, prepared, propagated, compounded, or processed in an establishment not so registered are imported or offered for import into the United States, samples of such drugs or devices be delivered to the Secretary of Health and Human Services, with notice of such delivery to the owner or consignee, who may appear before the Secretary of Health and Human Services and have the right to introduce testimony. If it appears from the examination of such samples or otherwise that (1) such article has been manufactured, processed, or packed under insanitary conditions or, in the case of a device, the methods used in, or the facilities or controls used for, the manufacture, packing, storage, or installation of the device do not conform to the requirements of section 520(f), or (2) such article is forbidden or restricted in sale in the country in which it was produced or from which it was exported, or (3) such article is adulterated, misbranded, or in violation of section 505, then such article shall be refused admission, except as provided in subsection (b) of this section. The Secretary of the Treasury shall cause the destruction of any such article refused admission unless such article is exported, under regulations prescribed by the Secretary of the Treasury, within ninety days of the date of notice of such refusal or within such additional time as may be permitted pursuant to such regulations. Clause (2) of the third sentence of this paragraph shall not be construed to prohibit the admission of narcotic drugs the importation of which is permitted under the Controlled Substances Import and Export Act.

(b) Pending decision as to the admission of an article being imported or offered for import, the Secretary of the Treasury may authorize delivery of such article to the owner or consignee upon the execution by him of a good and sufficient bond providing for the payment of such liquidated damages in the event of default as may be required pursuant to regulations of the Secretary of the Treasury. If it appears to the Secretary of Health and Human Services that an article included within the provisions of clause (3) of subsection (a) of this section can, by relabeling or other action, be brought into compliance with the Act or rendered other than a food, drug, device, or cosmetic, final determination as to admission of such article may be deferred and, upon filing of timely written application by the owner or consignee and the execution by him of a bond as provided in the preceding provisions of this subsection, the Secretary may, in accordance with regulations, authorize the applicant to perform such relabeling or other action specified in such authorization (including destruction or export of rejected articles or portions thereof, as may be specified in the Secretary's authorization). All such relabeling or other action pursuant to such authorization shall in accordance with regulations be under the supervision of an officer or employee of the Department of Health and Human Services designated by the Secretary, or an officer or employee of the Department of the Treasury designated by the Secretary of the Treasury.

(c) All expenses (including travel, per diem or subsistence, and salaries of officers or employees of the United States) in connection with the destruction provided for in subsection (a) of this section and the supervision of the relabeling or other action authorized under the provisions of subsection (b) of this section, the amount of such expenses to be determined in accordance with regulations, and all expenses in connection with the storage, cartage, or labor with respect to any article refused admission under subsection (a) of this section, shall be paid by the owner or consignee and, in default of such payment, shall constitute a lien against any future importations made by such owner or consignee.

(d)[70] (1) A food, drug, device, or cosmetic intended for export shall not be deemed to be adulterated or misbranded under this Act if it–

(A) accords to the specifications of the foreign purchaser,

(B) is not in conflict with the laws of the country to which it is intended for export,

(C) is labeled on the outside of the shipping package that it is intended for export, and

(D) is not sold or offered for sale in domestic commerce.

---

[70] Sec. 801(d) amended by sec. 106 of P.L. 90-399, and sec. 3(f) of P.L. 94-295.

This paragraph does not authorize the exportation of any new animal drug, or animal feed bearing or containing a new animal drug, which is unsafe within the meaning of section 512.

(2) Paragraph (1) does not apply to any device

(A) which does not comply with an applicable requirement of section 514 or 515,

(B) which under section 520(g) is exempt from either such section, or

(C) which is a banned device under section 516,

unless, in addition to the requirements of paragraph (1), the Secretary has determined that the exportation of the device is not contrary to public health and safety and has the approval of the country to which it is intended for export.

SECTION IX - MISCELLANEOUS

SEPARABILITY CLAUSE

SEC. 901.[391] If any provision of this Act is declared unconstitutional, or the applicability thereof to any person or circumstances is held invalid, the constitutionality of the remainder of the Act and the applicability thereof to other persons and circumstances shall not be affected thereby.

EFFECTIVE DATE AND REPEALS

SEC. 902.[392]

(a) This act shall take effect twelve months after the date of its enactment. The Federal Food and Drug Act of June 30, 1906, as amended (U.S.C., 1934 ed., title 21, secs. 1-15), shall remain in force until such effective date, and except as otherwise provided in this subsection, is hereby repealed effective upon such date: *Provided,* That the provisions of section 701 shall become effective on the enactment of this Act, and thereafter the Secretary of Agriculture is authorized hereby to (1) conduct hearings and to promulgate regulations which shall become effective on or after the effective date of this Act as the Secretary of Agriculture shall direct, and (2) designate prior to the effective date of this Act food having common or usual names and exempt such food from the requirements of clause (2) of section 403(i) for a reasonable time to permit the formulation, promulgation, and effective application of definitions and standards of identity therefor as provided by section 401: *Provided further,* That sections (502(j), 505, and 601(a), and all other provisions of this Act to the extent that they may relate to the enforcement of such sections, shall take effect on the date of the enactment of this Act, except that in the case of a cosmetic to which the proviso of section 601(a) relates, such cosmetic shall not, prior to the ninetieth day after such date of enactment, be deemed adulterated by reason of the failure of its label to bear the legend prescribed in such proviso: *Provided further,* That the Act of March 4, 1923 (U.S.C., 1945 ed., title 21, sec. 321a; 32 Stat. 1500, ch. 268), defining butter and providing a standard therefor; the Act of July 24, 1919 (U.S.C., 1946 ed., (title 21, sec. 321b; 41 Stat. 271, ch. 26), defining wrapped meats[71] as in package form; and the amendment to the Food and Drug Act, section 10A, approved August 27, 1935 (U.S.C., 1946 ed., title 21, sec. 372a[72] [49 Stat. 871, ch. 739]), shall remain in force and effect and be applicable to the provisions of this Act.

(b) Meats and meat food products shall be exempt from the provisions of this Act to the extent of the application or the extension thereto of the Meat Inspection Act, approved March 4, 1907, as amended (U.S.C., 1946 ed., title 21, secs. 71-96; 34 Stat. 1260 *et seq.*).

[SEC. 7. Public Law 85-929 (21 U.S.C. 451 note): Nothing in this Act shall be construed to exempt any meat or meat food product or any person from any requirement imposed by or pursuant to the Poultry Products Inspection Act (21 U.S.C. 451 and the following) or the Meat Inspection Act of March 4, 1907, 34 Stat. 1260, as amended and extended (21 U.S.C. 71 and the following).]

---

[71] See secs. 201a and 201b.

[72] See footnote 31.

(c)[73] Nothing contained in this Act shall be construed as in any way affecting, modifying, repealing, or superseding the provisions of section 351 of Public Health Service Act (relating to viruses, serums, toxins, and analogous products applicable to man); the virus, serum, toxin, and analogous products provisions, applicable to domestic animals, of the Act of Congress approved March 4, 1913 (37 Stat. 832-833); the Filled Cheese Act of June 6, 1896 (U.S.C., 1946 ed., title 26, ch. 17, secs. 2350-2362); the Filled Milk Act of March 4, 1923 (U.S.C., 1946 ed., title 21, ch. 3, secs. 61-64); or the Import Milk Act of February 15, 1927 (U.S.C., 1946 ed., title 21, ch. 4, secs. 141-149).

(Approved June 25, 1938.)

[Excerpt from P.L. 88-136, October 11, 1963]

REVOLVING FUND FOR CERTIFICATION AND OTHER SERVICES

For the establishment of a revolving fund for certification and other services, there is hereby appropriated the aggregate of fees (including advance deposits to such fees) paid during the fiscal year 1964, and each succeeding fiscal year, for services in connection with the listing, certification, or inspection of certain products and the establishment of tolerances for pesticides, in accordance with sections 406, 408, 506, 507, 702A, and 706 of the Federal Food, Drug, and Cosmetic Act, as amended (21 U.S.C. 346a, 356, 357, 372a, and 376), and the unexpended balance of such fees (or advance deposits) heretofore appropriated shall be credited to such revolving fund. This fund shall be available without fiscal year limitation for salaries and expenses necessary to carry out the Secretary's responsibilities in connection with such listings, certifications, inspections, or establishment of tolerances, including the conduct of scientific research, development of methods of analysis, purchase of

[73] Sec. 902(c)amended by sec. 107 of P.L. 90-399. Nothing in the amendments made by the Drug Amendments of 1962 to the Federal Food, Drug, and Cosmetic Act shall be construed as invalidating any provision of State law which would be valid in the absence of such amendments unless there is a direct and positive conflict between such amendments and such provision of State law.

chemicals, fixtures, furniture, and scientific equipment and apparatus; expenses of advisory committees; refund of advance deposits for which no services have been rendered: *Provided,* That any supplies, furniture, fixtures, and equipment on hand or on order on June 30, 1963, and purchased or ordered under appropriations for "Salaries and Expenses, Certification, Inspection, and Other Services," shall be used to capitalize the revolving fund.

CGMP'S/FOOD PLANT SANITATION 2ND Edition

SUGGESTED BASIC REFERENCE TEXTS

ANON, 1968. Laboratory Manual for Food Canners and Processors, 2 Vol., National Food Processors Assn., Washington, D.C.

ANON, 1974. Guidelines for Product Recall. Grocery Manufacturers of America, Inc., Washington, D.C.

ANON, 1985. FDA Compliance Program Guidance Manual. Section I Foods and Cosmetics Basic Section for FY 86., US Dept. of Health and Human Services, Public Health Service, Food and Drug Administration. Reproduced by National Technical Service, U.S. Dept. of Commerce, Springfield, VA.

ANON, 1986. FDA Inspection Operations Manual. Dept of Health and Human Services, Public Health Service, Food and Drug Administration. Reproduced by National Technical Information Service, U.S. Dept. of Commerce, Springfield, VA.

BANWART, George J., 1979, Basic Food Microbiology. AVI Publishing Company, Inc. Westport, CT.

GOULD, Wilbur A. 1977. Good Manufacturing Practices for Snack Food Manufacturers. Published by Potato Chip/Snack Food Association. Alexandria, Va.

GOULD, Wilbur A. and Ronald W. Gould, 1988. Total Quality Assurance For The Food Industries. CTI Publications, Inc., Baltimore, MD.

IMHOLT, Thomas J., 1984. Engineering for Food Safety and Sanitation. Technical Institute of Food Safety, Crystal, MN.

SUGGESTED BASIC REFERENCE TEXTS - Continued

KATSUYAMA, Allen M. and Jill P. Strachan, 1980. Principles of Food Processing Sanitation. The Food Processors Institute, Washington, D.C.

LOPEZ, Anthony, 1987. A Complete Course in Canning, 3 Volumes. Published by The Canning Trade Inc., Baltimore, MD.

MOUNTNEY, George J. and Wilbur A. Gould, 1988. Practical Food Microbiology, Van Nostrand Reinhold Company, New York, NY.

PARKER, Milton E. and John H. Litchfield, 1962. Food Plant Sanitation. Reinhold Publishing Company, New York, NY.

FIGURES

TABLES

FORMS

CHARTS

NOTES

NOTES

NOTES

NOTES

NOTES

NOTES

NOTES

NOTES

NOTES

NOTES

NOTES

NOTES